A TEACHER'S GUIDE TO
Using the
Next Generation
Science
Standards With Gifted
and Advanced Learners

A TEACHER'S GUIDE TO
Using the
Next Generation
Science
Standards With Gifted
and Advanced Learners

Cheryll M. Adams, Ph.D.,
Alicia Cotabish, Ed.D.,
and Debbie Dailey, Ed.D.

A Service Publication of the

NATIONAL ASSOCIATION FOR
Gifted Children

Routledge
Taylor & Francis Group

NEW YORK AND LONDON

Library of Congress Cataloging-in-Publication Data

Adams, Cheryll M., 1948-
 A teacher's guide to using the next generation science standards with gifted and advanced learners / by Cheryll M. Adams, Ph.D., Alicia Cotabish, Ed. D., & Debbie Dailey, Ed.D.
     pages cm
 Includes bibliographical references.
 ISBN 978-1-61821-283-2 (pbk.)
 1. Science--Study and teaching--United States. 2. Science--Study and teaching--Standards--United States. 3. Gifted children--Education--United States. 4. Gifted children--Identification. I. Cotabish, Alicia. II. Dailey, Debbie, 1963- III. Title.
 Q183.3.A1A338 2014
 507.1'273--dc23

                                        2014024583

First published in 2015 by Prufrock Press Inc.

Published in 2021 by Routledge
605 Third Avenue, New York, NY 10017
2 Park Square, Milton Park, Abingdon, Oxon OX14 4RN

Routledge is an imprint of the Taylor & Francis Group, an informa business

Copyright ©2015 National Association for Gifted Children

Production design by Raquel Trevino

ISBN: 9781032143286(hbk)
ISBN: 9781618212832(pbk)

DOI: 10.4324/9781003238522

Table of Contents

Acknowledgments

Many people have assisted with the efforts in developing this book. They include the leadership of the National Association for Gifted Children (NAGC), the NAGC Professional Standards Committee, reviewers, NAGC staff, and experts who were a part of the development of the other books in this series on using the Common Core State Standards (CCSS) with gifted and advanced learners.

We would like to thank Tracy L. Cross, NAGC president; past president Paula Olszewski–Kubilius; and the NAGC Board, who have understood the urgency for responding to the national standards movement, including the Next Generation Science Standards (NGSS) and the Common Core State Standards, and the gifted education community's need to have a voice in their implementation. From the beginning, the NAGC Professional Standards Committee also has been actively involved in providing the framework, expertise, and support for this book. Moreover, the NAGC leadership group also includes Executive Director Nancy Green and Association Editor Carolyn Callahan, who have both supported the development process and the need for this book.

This book has also been strengthened through a rigorous review process. We want to thank the reviewers Steve Coxon, Jeff Danielian, and Chrys Mursky, who took time to provide valuable advice and feedback.

Finally, the authors want to express a special thank you to Jane Clarenbach, Director of Public Education at the NAGC office, who has provided the needed energy in supporting the authors through the process and a critical eye in editing the many drafts of this book.

Foreword

A Teacher's Guide to Using the Next Generation Science Standards With Gifted and Advanced Learners is the second publication from the National Association for Gifted Children (NAGC) on differentiating the Next Generation Science Standards (NGSS). Although the NGSS to date have not been as widely adopted as the Common Core State Standards (CCSS), they provide the opportunity to bring gifted education strategies to all students, not only to those who have been identified as gifted. Talent development based on these strategies begins early, and users of this book will find examples of challenging learning experiences as early as kindergarten. In an effort to facilitate advanced learning as it relates to these important standards, NAGC's goal is to help teachers encourage early scientific thinking as a way to identify science talent. In addition, it has much to offer that will help teachers to develop confidence to modify and extend the standards for students ready to move at a faster pace or who are interested in an in-depth examination of a core idea within the standards.

To be effective with NGSS, early education teachers will need more knowledge of science content and secondary teachers will need additional pedagogical skills than either might have received in their teacher preparation programs. Thus, professional development opportunities will be essential in successfully implementing the standards with all students. The authors, who are well qualified in gifted education strategies, describe the elements of professional development that will ensure teachers are able to make the needed changes in their classrooms to challenge and support advanced learners.

There is much to celebrate with increased emphasis on rigorous standards, but we must be careful not to think that gifted and advanced learners will be appropriately challenged without additional work on the part of schools. I am confident that this book will be a useful resource for all educators, not just those working specifically in gifted education programs.

As is often the case, NAGC is indebted to a group of experts who volunteered their time and expertise to ensure that members and others who work with advanced and gifted students in science have outstanding resources that support teaching and learning related to these important new standards. We thank Cheryll Adams, Alicia Cotabish, and Debbie Dailey for their contributions to this project.

Tracy L. Cross, Ph.D.
President, NAGC

Chapter 1

Overview

The purpose of this book is to provide classroom teachers and school administrators examples and strategies to implement the new Next Generation Science Standards (NGSS) for advanced learners at all stages of development. One aspect of fulfilling that purpose is to clarify what advanced opportunities look like for such learners as they progress from kindergarten through high school. How can teachers provide the level of rigor and relevance within the new standards as they translate them into experiences for gifted learners? How can they provide creative and innovative opportunities that will nurture the thinking and problem solving of our best students?

This book also serves as a primer for guiding policies and practices related to advanced learners in school. At all levels, schools must be flexible in the implementation of policies related to acceleration, waivers, and course credit, all of which may impact gifted learners. The developers of the NGSS acknowledge that advanced learners may move through the standards more readily than other learners (Achieve, Inc., 2014a), attesting to the importance of using differentiated approaches for these learners to attain mastery and/or progress in academic achievement at

DOI: 10.4324/9781003238522-1

their level. It is critical that schools allow for flexibility in these areas and others in order to accommodate the special needs of our advanced learners.

This book is based on a set of underlying assumptions about the constructs of giftedness and talent development that underpin the thinking that spawned *Using the Next Generation Science Standards With Gifted and Advanced Learners* (Adams, Cotabish, & Ricci, 2014). These assumptions are:

- Giftedness is developed over time through the interaction of innate abilities with nurturing environmental conditions. Thus the process is developmental, dynamic, and malleable.

- Many learners show preferences for particular subject matter early and continue to select learning opportunities that match their predispositions if they are provided with opportunities to do so. For many children, especially those in poverty, schools are the primary source for relevant opportunities to develop domain-specific potential, although markers of talent development also emerge from work done outside of school in cocurricular or extracurricular contexts.

- Aptitudes may emerge as a result of exposure to high-level, challenging activities in an area of interest. Thus teachers should consider using advanced learning activities and techniques as a stimulus for all learners.

- In the talent development process, there is an interaction effect between affect and cognition, leading to heightened intrinsic motivation of the individual and focus on the enjoyable tasks associated with the talent area. This dynamic tension catalyzes movement to the next level of advanced work in the area.

- Intellectual cultural diversity among students may account for different rates of learning, areas of aptitude, cognitive styles, and experiential backgrounds. If they are to meet such diverse student needs, teachers should differentiate and customize curriculum and instruction, always working to provide an optimal match between the learner and her readiness to encounter the next level of challenge.

Users of this book need to be sensitive to the ideas contained herein as not intended to apply exclusively to identified gifted students. Many gifted children go unidentified, especially if they are culturally diverse and/or from low socioeconomic status groups. The book also applies to students with potential in science, as they might develop motivation and readiness to learn within the domain of science.

Finally, it is our hope that the book provides a roadmap for meaningful national, state, and local educational reform that elevates learning in science to higher levels of passion, proficiency, and creativity for all learners.

The Next Generation Science Standards: What Are They?

The NGSS are standards for K–12 science education illustrating the curriculum emphases needed for students to develop scientific literacy for college readiness and the 21st century. Based on *A Framework for K–12 Science Education: Practices, Crosscutting Concepts, and Core Ideas* (National Research Council [NRC], 2012) and developed by experts across the disciplines of science, engineering, cognitive science, teaching and learning, curriculum, assessment, and education policy, the evolution of the NGSS included having the scientific and educational research communities identify core ideas in science and articulate them across grade bands (grades K–2, grades 3–5, grades 6–8, and grades 9–12). In the development phase of the standards, 26 states provided leadership by addressing common issues involved in adoption and implementation of the standards. The initiative was coordinated by Achieve, Inc., a nonprofit bipartisan organization, and involved a range of networks including the 35-state

American Diploma Project Network (ADPN) and the network of 24 states in the Partnership for Assessment of Readiness for College and Careers (PARCC). The state-led process of development included state policy leaders, higher education professionals, K–12 teachers, and the science and business communities.

When navigating the standards, educators have two options to view the standards: by topical arrangement (much like the arrangement of most standards in education) or by Disciplinary Core Ideas (physical science; life science; Earth and space science; engineering, technology, and applications of science). Furthermore, users can easily navigate the standards by topical arrangement or disciplinary core idea through an interactive filtering system available on the NGSS website (http://www.nextgenscience.org/next-generation-science-standards).

Three Dimensions of the Next Generation Science Standards

The NGSS authors combined three dimensions of science to form each standard. The dimensions encompass a vision of what it means to be a scientist. Educators should be aware of the following dimensions as they plan to work with the standards:

- *Dimension 1: Science and Engineering Practices* describes behaviors of scientists, explains and extends what is meant by "inquiry" in science, and focuses on the knowledge beyond skills that is needed to engage in science.
- *Dimension 2: Crosscutting Concepts* cohesively links different concepts of science that have application across domains. They include: Patterns, Similarity, and Diversity; Causes and Effect; Scale, Proportion, and Quantity; Systems and System Models; Energy and Matter; Structure and Functions; and Stability and Change.
- *Dimension 3: Disciplinary Core Ideas* (DCI) is grouped in four domains: the physical sciences; the life sciences; the

Earth and space sciences; and engineering, technology, and applications of science. DCI are grounded in K–12 science curriculum, instruction, and assessment, and are considered to be the most important aspects in the teaching and learning of science. DCI are shaped by ideas that have broad discipline importance, key organizing concepts, key features of understanding or investigating complex ideas in science, and student and societal impact. The following sections provide more information about each.

Dimension 1: Science and Engineering Practices

When considering the implications of the NGSS for the development of science talent, it is important to take into account the eight standards for science and engineering practices that educators should seek to develop in their students, as well as the individual science content standards. According to the authors of the NGSS, these practices describe behaviors that scientists engage in "as they investigate and build models and theories about the natural world and the key set of engineering practices that engineers use as they design and build models and systems" (Achieve, Inc., 2014b, para. 2). The scientific and engineering practices are an integral part of the NGSS. They build on the NRC's (2012) *Framework for K–12 Education*, produced for the NGSS. The practices increase in complexity and sophistication across grade levels and are intended for use with all students from kindergarten through college and careers:

1. Asking questions (for science) and defining problems (for engineering)
2. Developing and using models
3. Planning and carrying out investigations
4. Analyzing and interpreting data
5. Using mathematics and computational thinking
6. Constructing explanations (for science) and design solutions (for engineering)
7. Engaging in argument from evidence

8. Obtaining, evaluating, and communicating information

It is important that students actively engage in these practices daily in their science classes. Students need ongoing opportunities to experience the joy of investigating rich concepts in depth and applying reasoning and justification to a variety of scientific, engineering, and other problems.

In response to the release of the Common Core State Standards (CCSS) for Mathematics, Johnsen and Sheffield (2013) proposed a ninth Standard for Mathematical Practice focused on creativity and innovation. Given how vital this is for the 21st century, we proposed that a ninth Science and Engineering Practice be added for the development of promising science students:

9. Solving problems in novel ways and posing new scientific questions of interest to investigate

With our proposed standard, students are encouraged and supported in taking risks, embracing challenge, solving problems in a variety of ways, posing new scientific questions of interest to investigate, and being passionate about scientific investigations.

Dimension 2: Crosscutting Concepts

The NGSS Crosscutting Concepts are application-based concepts that cut across multiple domains of science. The concepts are an organizational schema for interrelating knowledge and represent a more integrated view of science learning. Specifically, the Crosscutting Concepts are:

1. Patterns
2. Cause and Effect: Mechanism and Explanation
3. Scale, Proportion, and Quantity
4. Systems and System Models
5. Energy and Matter: Flow, Cycle, and Conservation
6. Structure and Function
7. Stability and Change

Crosscutting Concepts are arranged in grade bands, which lessen ceiling effects by allowing students to explore concepts through multiple avenues. They represent conceptual characteristics of what scientists should be able to "do" and cut across multiple domains. For example, consider a task to explain the effect mass has on a falling object. It could be assessed using the grade band information described in the Cause and Effect: Mechanisms and Explanation concept. The expectation could be for students to use conceptual models (e.g., Newton's Second Law of Motion) in concert with a practice, such as modeling, to develop a structure or function (using different materials) to demonstrate the effects of mass on a falling object. Related tasks could be planning or carrying out investigations using mathematical and computational thinking, which are both part of the Science and Engineering Practices dimension. The tasks can be conducted over time to develop a portfolio of evidence about students' understandings and enactments of Crosscutting Concepts. For the gifted learner, advanced and complex tasks should be integrated to elevate learning.

Dimension 3: Disciplinary Core Ideas

Disciplinary Core Ideas (DCI) demonstrate a progression of ideas arranged in grade bands across four domains: the physical sciences; the life sciences; the earth and space sciences; and engineering, technology, and applications of science. To be considered core in the NGSS, the ideas have to meet at least two of the following criteria:

1. Have broad importance across multiple sciences or engineering disciplines or be a key organizing concept of a single discipline.
2. Provide a key tool for understanding or investigating more complex ideas and solving problems.
3. Relate to the interests and life experiences of students or be connected to societal or personal concerns that require scientific or technological knowledge.

4. Be teachable and learnable over multiple grades at increasing levels of depth and complexity.

The organization of the DCI into grade bands creates overlapping concepts at times; however, the arrangement is conducive to an integrated approach to science learning and allows for an accelerated trajectory for gifted learners.

Alignment of the NGSS
With Other Standards

All differentiation for the gifted is based on an understanding of the characteristics of gifted and high-potential students, the content standards within a domain, and the process of scientific inquiry. The NGSS provide an opportunity for the field of gifted education to examine its practices and align them more fully to the *NAGC Pre-K–Grade 12 Gifted Programming Standards* (NAGC, 2010) for curriculum, instruction, and assessment. For example, similar to the NAGC Programming Standards, which represent the professional standards for programs in gifted education across P–12 levels, the NGSS emphasize problem solving (NAGC, 2010) and the NGSS Disciplinary Core Ideas spread across the four domains of (a) physical sciences, (b) life sciences, (c) Earth and space sciences, and (d) engineering, technology, and applications of sciences. Because the gifted programming standards in curriculum require educators to engage in two major tasks in curriculum planning—alignment to standards in the content areas and the development of a scope and sequence—using the NGSS is a natural point of departure. The effort must occur in vertical planning teams within districts and states in order to increase the likelihood of consistency and coherence in the process.

Within the gifted education programming standards, the curriculum and assessment standards were used to design this book in the following ways:

- *Development of scope and sequence.* The authors have demonstrated a set of interrelated emphases and activities for use across K–12 with a common format and within key content domains.
- *Use of differentiation strategies.* The authors have used the central differentiation strategies emphasized in the national P–12 gifted programming standards, including critical and creative thinking, problem solving, inquiry, research, and conceptual development.
- *Use of appropriate pacing/acceleration techniques.* The authors used all of these strategies, as well as more advanced, innovative, and complex science learning experiences to ensure the challenge level for gifted learners.
- *Adaptation or replacement of the core curriculum.* Adaptation or replacement of the core curriculum extends the NGSS by ensuring that advanced and gifted learners master them and then go beyond them in key ways. Some standards may be mastered earlier and the science and engineering practices should be used consistently throughout the curriculum.
- *Use of research-based materials.* The authors have included research-based materials found to be highly effective with advanced and gifted learners in enhancing critical thinking, reasoning and sense making, problem solving, and innovation. These are included in Appendix C.
- *Use of information technologies.* The learning experiences are easily adaptable for use with multimedia technologies.
- *Use of metacognitive strategies.* The authors included activities where students use reflection, planning, monitoring, and assessing skills.
- *Talent development in areas of aptitude and interest in various domains (e.g., cognitive, affective, aesthetic).* The book presents examples that provide multiple opportunities for students to explore domain-specific interests, such as conducting research, investigating problems, creating models, and exercising multiple levels of skills in cogni-

tive, affective, and aesthetic areas with special attention given to the integration of engineering practices. The book also presents differentiated learning examples for problem- and project-based learning activities.

21st-Century Skills

This book also includes a major emphasis on key 21st-century skills (Partnership for 21st Century Skills, 2009) in overall orientation as well as in the instructional experiences and assessments employed in the examples. The National Science Teachers Association (NSTA) recognized the important connection between science education and 21st-century skills, stating, "Exemplary science education can offer a rich context for developing many 21st century skills, such as critical thinking, problem solving, and information literacy, especially when instruction addresses the nature of science and promotes use of science practices" (NSTA, 2011, para. 4). Through the exemplars presented in this book, students will consistently think critically, communicate ideas and findings, and collaborate on hypotheses, experiments, and data collection.

Although many 21st-century skills are embedded in the sample learning experiences for both the typical and advanced learner, all challenging science instruction should be supported by the following 21st-century skills.

- *Collaboration*: Students are encouraged to work with partners and small groups to carry out tasks and projects, to pose and solve problems, and to plan presentations (Partnership for 21st Century Skills, 2009).
- *Communication*: Students are encouraged to develop communication skills in written, oral, visual, and technological modes in a balanced format within each unit of study (Partnership for 21st Century Skills, 2009).
- *Critical thinking*: Students are provided with models of critical thought that are incorporated into classroom

instructional experiences, questions, and assignments (Partnership for 21st Century Skills, 2009).

- *Creative thinking*: Students are provided opportunities to think creatively so that they can develop skills that support original, innovative thinking, elaboration of ideas, flexibility of thought, and problem posing and solving (Partnership for 21st Century Skills, 2009).
- *Problem solving*: Students are engaged in real-world problem solving embedded in scientific processes in sample learning tasks.
- *Technology literacy*: Students use technology in multiple forms and formats as a tool in solving problems and to create generative products.
- *Information media literacy*: Students use multimedia to express ideas, research results, explore real-world problems, and evaluate information presented in media (graphs and diagrams) for scientific accuracy.
- *Social skills*: Students work in small groups and develop the tools of collaboration, communication, and working effectively with others on a common set of tasks.

To visually demonstrate the relationship between the NGSS and other standards and skills, student practices were examined. Figure 1.1 highlights some of the relationships between the NAGC Pre-K–Grade 12 Gifted Programming Standards, the NGSS (Science and Engineering Practices), and the CCSS in English language arts (student portraits) and Mathematics. Practices and portraits were grouped to illustrate student-centered expectations.

The midpoint of the graphic demonstrates the relationship across student-centered expectations and/or similar tenets among the four sets of standards. Furthermore, standard-specific student expectations that do not overlap across the four standards are listed in separate boxes. Please note that the graphic does not account for overlapping that may occur among two or three stan-

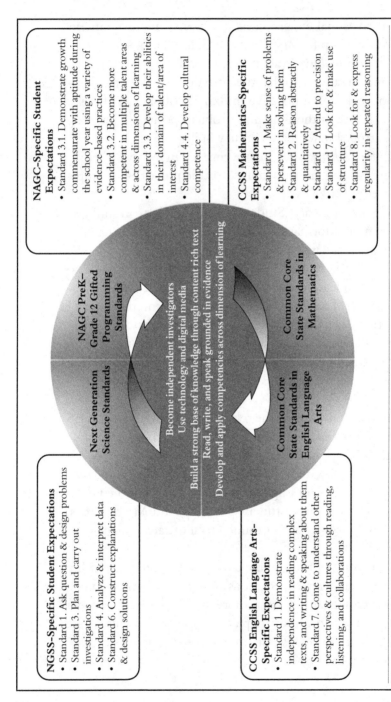

NAGC-Specific Student Expectations
- Standard 3.1. Demonstrate growth commensurate with aptitude during the school year using a variety of evidence-based practices
- Standard 3.2. Become more competent in multiple talent areas & across dimensions of learning
- Standard 3.3. Develop their abilities in their domain of talent/area of interest
- Standard 4.4. Develop cultural competence

CCSS Mathematics-Specific Expectations
- Standard 1. Make sense of problems & persevere in solving them
- Standard 2. Reason abstractly & quantitatively
- Standard 6. Attend to precision
- Standard 7. Look for & make use of structure
- Standard 8. Look for & express regularity in repeated reasoning

NGSS-Specific Student Expectations
- Standard 1. Ask question & design problems
- Standard 3. Plan and carry out investigations
- Standard 4. Analyze & interpret data
- Standard 6. Construct explanations & design solutions

CCSS English Language Arts–Specific Expectations
- Standard 1. Demonstrate independence in reading complex texts, and writing & speaking about them
- Standard 7. Come to understand other perspectives & cultures through reading, listening, and collaborations

NAGC PreK–Grade 12 Gifted Programming Standards

Next Generation Science Standards

Common Core State Standards in Mathematics

Common Core State Standards in English Language Arts

Become independent investigators
Use technology and digital media
Build a strong base of knowledge through content rich text
Read, write, and speak grounded in evidence
Develop and apply competencies across dimension of learning

Figure 1.1. Relationships and convergences found in the NGSS, the Common Core State Standards in English Language Arts, and the NAGC Pre-K–Grade 12 Gifted Programming standards. From *Relationships and Convergences Found in Common Core State Standards in Mathematics, Common Core State Standards in ELA/Literacy, and A Framework for K–12 Science Education,* by T. Cheuk, 2012, Arlington, VA: National Science Teachers Association. Copyright 2012 by NSTA. Adapted with permission.

dards (e.g., the use of mathematical and computational thinking in both mathematics and science).

Linking the NGSS to CCSS Math and English Language Arts

Because standards often can be addressed across subject areas rather than only in one domain, examples of how to link the NGSS to the Common Core State Standards (CCSS) in mathematics and English language arts (ELA) are included in each differentiated learning experience example (see Chapter 3). Other areas of learning that can be applied to standards-based tasks illustrate the efficiency and effectiveness that can be achieved through such compression and the differentiation for gifted learners that results. Dimension 2 of the NGSS, Crosscutting Concepts, links concepts such as patterns, cause and effect, and systems with each standard. Some of these concepts are also found in the CCSS, making it easier to find points of intersection. Furthermore, specific linkages to the CCSS in math and ELA are provided for each grade-level topic within the NGSS.

There are two ways to remodel content to engage and motivate highly able learners by making cross-disciplinary connections. Although the strategies are related, they are distinct. The first approach is to use cross-disciplinary content. The second is to integrate standards from science, English language arts, mathematics, or other disciplines. The following are a few examples for each strategy.

Integrating Standards

This strategy combines standards from two or more disciplines to increase complexity. For example:

- Using 1–LS1 (NGSS Life Science), From Molecules to Organisms: Structures and Processes: *Use materials to design a solution to a human problem by mimicking how plants and/or animals use their external parts to help them survive, grow, and meet their needs*; MP.4 (CCSS Mathematical Practice): *Model with mathematics*; and RI.2.1 (CCSS Reading Informational Texts): *Ask and answer such questions as* who, what, where, when, why, *and* how *to demonstrate understanding of key details in a text.* Ask students to design, demonstrate, graph, and explain (using who, what, where, when, why, and how) a product for humans that can protect them from predators (simulating plant and/or animal survival parts).

- Using 4–ESS2–1 (NGSS Earth Systems): *Make observations and/or measurements to provide evidence of the effects of weathering or the rate of erosion by water, ice, wind, or vegetation*; MP.2 (CCSS Mathematical Practice): *Reason abstractly, and quantitatively*; MP.4 (CCSS Mathematical Practice): *Model with mathematics*; and W.4.8 (CCSS ELA Writing): *Recall relevant information from experience from print and digital sources.* Students can observe, demonstrate, and measure the effects of weathering and erosion, and utilize digital resources to research how physical processes (e.g., climate, plate tectonics, erosion, soil formation, water cycle, and circulation patterns in the ocean shape patterns) influence the environment and the availability and quality of natural resources.

Cross-disciplinary and integrated approaches are inherent in many research projects that students undertake in science. The writing demonstrates the capacity to build argument, and the construction of mathematical models and analysis of data illustrate the capacity to interpret and transform ideas from graphic representations to verbal ones. By doing so, science, English language arts, and mathematics standards are addressed.

Finding and Developing Talent With NGSS and Gifted Education Strategies

To find science talent, teachers will need to observe students engaged in challenging, hands-on, inquiry-based activities, such as those that can be designed using all elements of the NGSS and differentiated for gifted and advanced learners: Disciplinary Core Ideas, Crosscutting Concepts, and Science and Engineering Practices, as well as links to the CCSS in math and ELA.

To nurture the science innovators of tomorrow, educators need to help students develop passion, perseverance, and creativity in the face of difficult problems and not just scientific competence in knowing facts and problem solving. Higher expectations are needed that include scientific creativity, which encourages students to create their own methods for making sense of and solving problems and raising new questions that are suggested during the solution of the original problem. The Science and Engineering Practices of the NGSS highlight this need.

When students with potential in science participate in accelerated classes that are taught by experienced teachers who are aware of their needs, they are more likely to take rigorous college courses, complete advanced degrees, and feel academically challenged and socially accepted (Colangelo, Assouline, & Gross, 2004; Gross, 2006; Rogers, 2007). Teachers and administrators must be aware of the challenges that gifted students may face when accelerating, particularly if they skip an entire academic year. Teachers can help gifted students to succeed by using formative assessment and other strategies to identify any gaps in knowledge (Chapman, 2009; Siegle & McCoach, 2002).

Other available options that provide academic challenge are mentorships, dual enrollment, internships, early entrance to college, independent study, and summer programs (Callahan & Kyburg, 2005; Olszewski–Kubilius, 2010; VanTassel–Baska, 2007). There have been some extracurricular programs that show promise, such as museum science programs (Melber, 2003) and afterschool science enrichment programs for girls (Wood, 2002).

Results from both programs indicate that students prefer inquiry-based science activities, exactly the kind of activity that is stressed in the NGSS.

The NGSS are designed to provide challenging, inquiry-based science learning experiences. By giving students opportunities to engage in interactive activities that require them to work like professionals in the field, teachers have daily occasions to "kid watch" and observe students who may be demonstrating gifted behaviors in science.

Developing Innovative and Creative Scientists

In the STEM areas, all students, including high-ability students, should be afforded the opportunity to develop their talent. To spur innovation in science, teachers should include an open-ended inquiry approach to teaching and learning, utilize higher order questioning, and provide accelerative learning opportunities to facilitate and advance learning in science. To assist students in their creation of new scientific insights, some suggested questions for creative scientific investigations are:

- **Who?** Who has another solution? Who has another method? Who agrees or disagrees?
- **What or what if?** What patterns do I see in this data? What generalizations might I make from the patterns? What proof do I have? What are the chances? What is the best answer, best method of solution, best strategy to begin with? What if I change one or more parts of the problem? What new problems might I create?
- **When?** When does this work? When does this not work?
- **Where?** Where did that come from? Where should I start? Where might I go next? Where might I find additional information?
- **Why or why not?** Why does that work? If it does not work, why does it not work?
- **How?** How is this like other scientific problems or patterns that I have seen? How does it differ? How does this relate to real-life situations or models? How many

solutions are possible? How many ways might I use it to represent, simulate, model, or visualize these ideas? How many ways might I sort, organize, and present this information? (Sheffield, 2006, pp. 1–11)

Stimulating Scientific Thinking

Teachers of the gifted should be mindful of the importance of providing inquiry and investigative skills and strategies to stimulate scientific reasoning and work with a wide range of scientific topics grounded in biology, chemistry, physics, and mathematics. Early exposure to topics such as probability, statistics, and logic are also viable approaches to be used to support applied and cross-curricular skills, including conducting meaningful research in science and engineering.

In encouraging these high levels of creativity and giftedness (Chapin, O'Connor, & Anderson, 2009), teachers should realize that the role of the student is to:
- think, reason, make sense, and go deeper;
- talk to a partner and generate new ideas;
- repeat and rephrase what others have said and explain why he or she agrees or disagrees;
- make generalizations and justify conclusions;
- add on new ideas, new methods of solution, new questions, and original problems and related solutions;
- record solutions, reasoning, and questions;
- pose new scientific questions of interest to investigate; and
- create innovative scientific problems and solutions.

The role of the teacher in supporting and facilitating these skills is to:
- ask questions that encourage scientific creativity, reasoning, and sense making;
- elicit, engage, and challenge each student's thinking;
- listen carefully to students' ideas;

- ask students to clarify, justify, connect, and extend their ideas;
- assist students in attaching mathematical notation and scientific language to their ideas;
- reflect on student understanding, differentiate instruction, and encourage participation; and
- guide students to resources, including online, in print, and in person such as mentors, apprenticeships, competitions, clubs, and other extracurricular opportunities.

Early exposure to scientific thinking and processes can stimulate innovation (National Science Board [NSB], 2010). Providing a challenging learning environment that includes advanced questioning, depth and complexity, and adjustments for instructional pace can facilitate a bridge between scientific knowledge, skills, and processes.

Chapter 2

Adapting Learning Progressions for Gifted and Advanced Learners

Because learning is an ongoing developmental progression and takes place over time, teachers need information designed to help them identify what is expected from their students and the key points along a path that indicate growth in a student's knowledge and skills. The NGSS are informed by learning progressions (LPs) that are used to describe the pathway that will lead students toward mastery of a scientific concept. Learning progressions are "descriptions of the successively more sophisticated ways of thinking about a topic that can follow and build on one another as children learn about and investigate a topic over a broad span of time (e.g., 6 to 8 years)" (NRC, 2007, p. 213). According to Duncan and Rivet (2013) learning progressions differ from traditional scope and sequences in two major ways. First, they are empirically based through research about how students actually learn science. Second, rather than simply "adding more information or details over time, LPs focus on deepening understandings and developing increased complexity, applicability, and epistemological rigor with each learning opportunity" (Duncan & Rivet, 2013, p. 396). The LPs have five essential components:

DOI: 10.4324/9781003238522-2

1. Learning targets or clear end points that are defined by societal aspirations and analysis of the central concepts and themes in a discipline;
2. Progress variables that identify the critical dimensions of understanding and skill that are being developed over time;
3. Levels of achievement or stages of progress that define significant intermediate steps in conceptual/skill development that most children might be expected to pass through on the path to attaining the desired proficiency;
4. Learning performances that are the operational definitions of what children's understanding and skills would look like at each of these stages of progress, and which provide the specifications for the development of assessments and activities that would locate where students are in their progress; and,
5. Assessments that measure student understanding of the key concepts or practices and can track their developmental progress over time. (NRC, 2007, p. 15)

Teachers' knowledge of LPs is important to their understanding of the articulation of the three dimensions of the NGSS (Disciplinary Core Ideas, Science and Engineering Practices, and Crosscutting Concepts) within and across grade levels so that their students develop a deeper, broader, and more sophisticated understanding of the "big ideas" of science (NRC, 2012). Tables 2.1 and 2.2 present the three dimensions of the NGSS and the Disciplinary Core Ideas with their corresponding sub-ideas. By examining each progression separately and across domains (the physical sciences; the life sciences; the Earth and space sciences; and engineering, technology, and applications of science), teachers are able to identify how performance expectations change and increase in rigor for the student, to see how they are connected to one another, and to establish learning goals so that each student is challenged and experiences continuous progress from elementary to middle to high school and to higher edu-

Table 2.1

The Three Dimensions of the NGSS Framework

1. Scientific and Engineering Practices (SEP)	2. Crosscutting Concepts (CC)
1. Asking questions (for science) and defining problems (for engineering)	1. Patterns
2. Developing and using models	2. Cause and effect: Mechanism and explanation
3. Planning and carrying out investigations	3. Scale, proportion, and quantity
4. Analyzing and interpreting data	4. Systems and system models
5. Using mathematics and computational thinking	5. Energy and matter: Flows, cycles, and conservation
6. Constructing explanations (for science) and designing solutions (for engineering)	6. Structure and function
7. Engaging in argument from evidence	7. Stability and change
8. Obtaining, evaluating, and communicating information	

3. Disciplinary Core Ideas (DCI)

Physical Sciences

PS 1: Matter and its interactions
PS 2: Motion and stability: Forces and interactions
PS 3: Energy
PS 4: Waves and their applications in technologies for information transfer

Life Sciences

LS 1: From molecules to organisms: Structures and processes
LS 2: Ecosystems: Interactions, energy, and dynamics
LS 3: Heredity: Inheritance and variation of traits
LS 4: Biological evolution: Unity and diversity

Earth and Space Sciences

ESS 1: Earth's place in the universe
ESS 2: Earth's systems
ESS 3: Earth and human activity

Engineering, Technology, and Applications of Science

ETS 1: Engineering design
ETS 2: Links among engineering, technology, science, and society

Note: Adapted from *A Framework for K–12 Science Education: Practices, Crosscutting Concepts, and Core Ideas* (p. 3), by National Research Council, 2012, Washington, DC: The National Academies Press. Copyright 2012 by National Research Council. Adapted with permission.

Table 2.2
Disciplinary Core Ideas (DCI) With Sub–Ideas

Physical Sciences

PS 1: Matter and its interactions
  PS1A: Structure and properties of matter
  PS1B: Chemical reactions
  PS1C: Nuclear processes
PS 2: Motion and stability: Forces and interactions
  PS2A: Forces and motion
  PS2B: Types of interactions
  PS2C: Stability and instability in physical systems
PS 3: Energy
  PS3A: Definition of energy
  PS3B: Conservation of energy and energy transfer
  PS3C: Relationship between energy and forces
  PS3D: Energy and chemical processes in everyday life
PS 4: Waves and their applications in technologies for information transfer
  PS4A: Wave properties
  PS4B: Electromagnetic radiation
  PS4C: Information technologies and instrumentation

Life Sciences

LS 1: From molecules to organisms: Structures and processes
  LS1A: Structure and function
  LS1B: Growth and development of organisms
  LS1C: Organization for matter and energy flow in organisms
  LS1D: Information processing
LS 2: Ecosystems: Interactions, energy, and dynamics
  LS2A: Interdependent relationships in ecosystems
  LS2B: Cycles of matter and energy transfer in ecosystems
  LS2C: Ecosystem dynamics, functioning, and resilience
  LS2D: Social interactions and group behavior
LS 3: Heredity: Inheritance and variation of traits
  LS3A: Inheritance of traits
  LS3B: Variation of traits
LS 4: Biological evolution: Unity and diversity
  LS4A: Evidence of common ancestry
  LS4B: Natural selection
  LS4C: Adaptation
  LS4D: Biodiversity and humans

Table 2.2, *continued*

Earth and Space Sciences	
ESS 1: Earth's place in the universe	ESS 3: Earth and human activity
ESS1A: The Universe and its stars	ESS3A: Natural resources
ESS1B: Earth and the solar system	ESS3B: Natural hazards
ESS1C: The history of planet Earth	ESS3C: Human impacts on Earth systems
ESS 2: Earth's systems	ESS3D: Global climate change
ESS2A: Earth materials and systems	
ESS2B: Plate tectonics and large scale system interactions	
ESS2C: The roles of water in Earth's surface processes	
ESS2D: Weather and climate	
ESS2E: Biogeology	
Engineering, Technology, and Applications of Science	
ETS 1: Engineering design	ETS 2: Links among engineering, technology, science, and society
ETS1A: Defining and delimiting engineering problems	
ETS1B: Developing possible solutions	
ETS1C: Optimizing the design solution	

Note: Adapted from *Next Generation Science Standards: For States, By States, Volume 1, The Standards* (p. xxiv), by NGSS Lead States, 2013a, Washington, DC: The National Academies Press. Copyright 2013 by The National Academies Press. Adapted with permission.

cation and professional careers. Learning progressions are also useful in designing formative and summative assessments and ensuring that above-level performance expectations are included so that gifted and advanced students' strengths in knowledge and skills can be identified.

Students who are gifted and advanced in science may proceed through LPs in a different sequence and at a different pace than their peers. For example, mapping backward from a set of desired outcomes such as attendance at a selective college or college entrance during high school may create a LP too steep for some students, yet not steep enough for others. Teachers need to be aware of not only the LPs for a typical learner but also for ways of adapting them for gifted and advanced learners. As the authors of the NGSS state, " . . . no one document can fully represent all of the interventions and support necessary for students with varying degrees of abilities and needs" (NGSS Lead States, 2013a, p. xvii). In addition, the NGSS are not an exhaustive list of every possible science topic that could be taught and "should not prevent students from going beyond the standards when appropriate" (p. xvii). This chapter will describe the NGSS LPs within and across dimensions and ways that they might be adapted for gifted and advanced learners.

Descriptions of the Learning Progressions

The NGSS have LPs in each of the NGSS dimensions that describe expectations within each grade band's endpoints, and expectations in earlier grades build a foundation for those in later grades. There are detailed and extensive LPs for each of the three dimensions in the appendices to the NGSS (NGSS Lead States, 2013b). Having each dimension's LPs listed separately may give teachers the impression that these dimensions are taught in isolation, but that would be a gross misconception. For each performance expectation in the DCIs, corresponding Science and Engineering Practices (SEP) and Crosscutting Concepts (CC) are provided. Thus the LPs provide an overview of what is expected at grade-band endpoints, and teachers will then need to go to the NGSS arranged by DCI or topic for guidance in teaching the three dimensions together. Both formats are useful for knowing when and how to differentiate the learning experiences for gifted and advanced science learners. Therefore, teachers need to know about the progressions within a dimension (DCI, SEP, or CC) or domain (the physical science; the life sciences; the Earth and space sciences; and engineering, technology, and applications of science) and across dimensions and domains so that major concepts and practices can be introduced, practiced, and extended. Figure 2.1 shows the components that make up each standard and how to locate them when looking at a page from the standards document (NGSS Lead States, 2013a).

Regardless of which dimension's LPs we are studying, there are particular characteristics that permeate all three. For example, concepts move from concrete in the early grades to abstract by high school. Observations and understandings begin at the macroscopic levels in kindergarten and early elementary grades, moving to microscopic, atomic, and cellular levels at middle school, and finally to subatomic and subcellular levels in high school. Direct observation and explanations of everyday experiences move through the grades toward explanations of abstract concepts and theories (NRC, 2012). Figure 2.2 shows the LP of the first DCI in life sciences; a partial list of DCIs can be found in Figures 2.3 and 2.4. The lefthand column contains statements of the disciplinary core idea (L1), while the rest of the boxes indicate what students should know, understand, and be able to do at the grade-band endpoints (grades 2, 5, 8, and 12). Notice how the learning becomes more sophisticated as we move from left to right within a row. For example, in the first row, LS1.A, primary students are concerned with external parts of organisms and how these parts are used in the organism's daily life, such as how the organism gets nourishment. By the time students leave high school, they have gone beyond macroscopic observations to studying systems of specialized cells within the organism, such as the different types of cells that make up the digestive

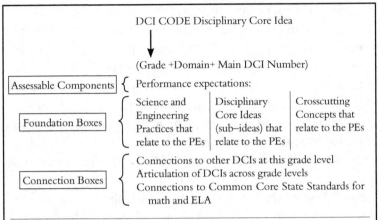

Figure 2.1. How to read the NGSS. Adapted from *Next Generation Science Standards: For States, By States, Volume 1, The Standards,* (pp. xxii–xxvi), by NGSS Lead States, 2013a, Washington, DC: The National Academies Press. Copyright 2013 by The National Academies Press. Adapted with permission.

system. The same process of building on prior knowledge and skills and greater sophistication of understanding can be seen in the Developing and Using Models LP from the SEP dimension and the Patterns LP from the Crosscutting Concepts Dimension. The scope of this chapter does not allow us to go into detail on every LP in each dimension, but charts similar to Figures 2.2–2.4 are found in Appendices E, F and G, respectively, of the NGSS (NGSS Lead States, 2013b). These charts are the starting points for differentiating for gifted and advanced learners in science.

Learning Progressions Within Domains

The NGSS identify LPs within each of the domains based on the logical progression of science, research about how students learn science, and research on children's cognitive development (NGSS Lead States, 2013a; NRC, 2012). For example, Table 2.3 on p. 32 shows how the concepts of force and motion become increasingly more sophisticated from kindergarten to high school within the physical sciences domain, DCI PS2. Note that in the NGSS, the concept is not taught in every grade.

Increasing Sophistication of Student Thinking	K–2	3–5	6–8	9–12
LS1.A Structure and function	All organisms have external parts that they use to perform daily functions.	Organisms have both internal and external macroscopic structures that allow for growth, survival, behavior, and reproduction.	All living things are made up of cells. In organisms, cells work together to form tissues and organs that are specialized for particular body functions.	Systems of specialized cells within organisms help perform essential functions of life. Any one system in an organism is made up of numerous parts. Feedback mechanisms maintain an organism's internal conditions within certain limits and mediate behaviors.
LS1.B Growth and development of organisms	Parents and offspring often engage in behaviors that help the offspring survive.	Reproduction is essential to every kind of organism. Organisms have unique and diverse life cycles. *continued*	Animals engage in behaviors that increase the odds of reproduction. An organism's growth is affected by both genetic and environmental factors.	Growth and division of cells in organisms occurs by mitosis and differentiation for specific cell types.
LS1.C Organization for matter and energy flow in organisms	Animals obtain food they need from plants or other animals. Plants need water and light.	Food provides animals with the materials and energy they need for body repair, growth, warmth, and motion. Plants acquire *continued*	Plants use the energy from light to make sugars through photosynthesis. Within individual organisms, food is broken down through a series of chemical reactions that *continued*	The hydrocarbon backbones of sugars produced through photosynthesis are used to make amino acids and other molecules that can be assembled into proteins or. *continued*

Figure 2.2. Partial life science Disciplinary Core Idea progressions in the NGSS. Adapted from *Next Generation Science Standards: Appendix E* (pp. 4–6), by NGSS Lead States, 2013c, Washington, DC: The National Academies Press. Copyright 2013 by The National Academies Press. Adapted with permission.

Figure 2.2, *continued*

Increasing Sophistication of Student Thinking	K–2	3–5	6–8	9–12
		material for growth chiefly from air, water, and process matter and obtain energy from sunlight, which is used to maintain conditions necessary for survival.	rearrange molecules and release energy.	DNA Through cellular respiration, matter and energy flow through different organizational levels of an organism as elements are recombined to form different products and transfer energy.
LS1.D Information Processing	Animals sense and communicate information and respond to inputs with behaviors that help them grow and survive.	Different sense receptors are specialized for particular kinds of information. Animals use their perceptions and memories to guide their actions.	Each sense receptor responds to different inputs, transmitting them as signals that travel along nerve cells to the brain: The signals are then processed in the brain, resulting in immediate behavior or memories.	N/A

Figure 2.2. *Continued.*

Grades K–2	Grades 3–5	Grades 6–8	Grades 9–12
Modeling in K–2 builds on prior experiences and progresses to include using and developing models (i.e., diagram, drawing, physical replica, diorama, dramatization, or storyboard) that represent concrete events or design solutions.	Modeling in 3–5 builds on K–2 experiences and progresses to building and revising simple models and using models to represent events and design solutions.	Modeling in 6–8 builds on K–5 experiences and progresses to developing, using, and revising models to describe, test, and predict more abstract phenomena and design systems.	Modeling in 9–12 builds on K–8 experiences and progresses to using, synthesizing, and developing models to predict and show relationships among variables between systems and their components in the natural and designed worlds.
• Distinguish between a model and the actual object, process, and/or events the model represents.	• Identify limitations of models.	• Evaluate limitations of a model for a proposed object or tool.	• Evaluate merits and limitations of two different models of the same proposed tool, process, mechanism or system in order to select or revise a model that best fits the evidence or design criteria.
• Compare models to identify common features and differences.	• Collaboratively develop and/or revise a model based on evidence that shows the relationships among variables for frequent and regular occurring events.	• Develop or modify a model—based on evidence – to match what happens if a variable or component of a system is changed.	
• Develop and/or use a model to represent amounts, relationships, relative scales (bigger, smaller); and/or patterns in the natural and designed world(s).	• Develop a model using an analogy, example, or abstract representation to describe a scientific principle or design solution.	• Use and/or develop a model of simple systems with uncertain and less predictable factors.	• Design a test of a model to ascertain its reliability.
• Develop a simple model based on evidence to represent a proposed object or tool.	• Develop and/or use models to describe and/or predict phenomena.	• Develop and/or revise a model to show the relationships among variables, including those that are not observable but predict observable phenomena.	• Develop, revise, and/or use a model based on evidence to illustrate and/or predict the relationships between systems or between components of a *continued*
	• Develop a diagram or simple *continued*	*continued*	

Figure 2.3. Partial Science and Engineering Practices progressions in the NGSS. Adapted from *Next Generation Science Standards: Appendix F* (p. 6), by NGSS Lead States, 2013d, Washington, DC: The National Academies Press. Copyright 2013 by The National Academies Press. Adapted with permission.

Figure 2.3, *continued*

Grades K–2	Grades 3–5	Grades 6–8	Grades 9–12
	physical prototype to convey a proposed object, tool, or process. • Use a model to test cause and effect relationships or interactions concerning the functioning of a natural or designed system.	• Develop and/or use a model to predict and/or describe phenomena. • Develop a model to describe unobservable mechanisms. • Develop and/or use a model to generate data to test ideas about phenomena in natural or designed systems, including those representing inputs and outputs, and those at unobservable scales.	system. • Develop and/or use multiple types of models to provide mechanistic accounts and/or predict phenomena, and move flexibly between model types based on merits and limitations. • Develop a complex model that allows for manipulation and testing of a proposed process or system. • Develop and/or use a model (including mathematical and computational) to generate data to support explanations, predict phenomena, analyze systems, and/or solve problems.

Figure 2.3. *Continued.*

1. Patterns—Observed patterns in nature guide organization and classification and prompt questions about relationships and causes underlying them.

K–2 Crosscutting Statements	3–5 Crosscutting Statements	6–8 Crosscutting Statements	9–12 Crosscutting Statements
• Patterns in the natural and human designed world can be observed, used to describe phenomena, and used as evidence.	• Similarities and differences in patterns can be used to sort, classify, communicate, and analyze simple rates of change for natural phenomena and designed products. • Patterns of change can be used to make predictions. • Patterns can be used as evidence to support an explanation.	• Macroscopic patterns are related to the nature of microscopic and atomic-level structure. • Patterns in rates of change and other numerical relationships can provide information about natural and human designed systems. • Patterns can be used to identify cause and effect relationships. • Graphs, charts, and images can be used to identify patterns in data.	• Different patterns may be observed at each of the scales at which a system is studied and can provide evidence for causality in explanations of phenomena. • Classifications or explanations used at one scale may fail or need revision when information from smaller or larger scales is introduced; thus requiring improved investigations and experiments. • Patterns of performance of designed systems can be analyzed and interpreted to reengineer and improve the system. • Mathematical representations are needed to identify some patterns. • Empirical evidence is needed to identify patterns.

Figure 2.4. Partial Crosscutting Concepts progressions in the NGSS. Adapted from *Next Generation Science Standards: Appendix G* (p. 15), by NGSS Lead States, 2013e, Washington, DC: The National Academies Press. Copyright 2013 by The National Academies Press. Adapted with permission.

Table 2.3

DCI PS2: Motion and Stability: Forces and Interactions.
Sub-Idea: PS2A: Force and Motion

Kindergarten	Grade 3	Middle School	High School
Pushes and pulls can have different strengths and directions.	Each force acts on one particular object and has both strength and direction.	For any pair of interacting objects, the force exerted by the first object on the second object is equal in strength to the force that the second object exerts on the first, but in the opposite direction. —(Newton's Third Law).	Newton's Second Law accurately predicts changes in the motion of macroscopic objects.

As mentioned in *A Framework for K–12 Science Education*, "Because research on these progressions is relatively recent, there is not a robust evidence base about appropriate sequencing for every concept, core idea, or practice identified in the framework" (NRC, 2012, p. 304). Hence, the NRC researchers caution that when evidence was not available, they consulted experts and used the best knowledge available at the time. Therefore, it is important that teachers identify performance expectations that meet the needs of individual students based on their previous knowledge and performance and their responsiveness to the curriculum and instruction.

Learning Progressions Across Dimensions

Recall that in the NGSS there are three dimensions, and although they are often discussed separately, the dimensions are integrated in the performance expectations (i.e., what students should be able to do). Thus, performance expectations must

> . . . [take into account that students cannot fully understand scientific and engineering ideas

without engaging in the practices of inquiry and the discourses by which such ideas are developed and refined]. At the same time, they cannot learn or show competence in practices except in the context of specific content. . . .Furthermore, crosscutting concepts have value because they provide students with connections and intellectual tools that are related across the differing areas of disciplinary content and can enrich their application of practices and their understanding of core ideas. (NRC 2012, p. 218)

In other words, students need to be able to apply the science and engineering practices and their understanding of the crosscutting concepts within the contexts of the four domains. Look back at Figure 1.1 again and notice that the information a teacher needs for any given performance expectation in any given domain is on the same page. Thus teachers will not need to search among multiple pages to find the appropriate DCI, SEP, and CC for any given performance expectation. The manner in which the NGSS are formatted underscores the necessity to integrate all three dimensions. Figure 2.5 shows an example of a PE with its associated DCI, CC, and SEP. Although only one grade's PE is illustrated here, each of the grades and grade bands has similar connections across dimensions. The understanding of how the dimensions relate to one another is essential in building the breadth and depth needed for designing rich learning experiences for gifted and advanced students. Again, the LPs can be differentiated within each of the domains by accelerating and combining student outcomes and across the dimensions by adding more complexity to the topic being discussed.

In summary, the LPs within domains and dimensions build upon and are linked to one another across grade levels. Looking at LPs across domains and dimensions provides greater clarity for incorporating more depth and complexity in learning experiences and in ways of compacting the standards for gifted and

MS-ESS1 Earth's Place in the Universe
MS-ESS1-1. Develop and use a model of the Earth-sun-moon system to describe the cyclic patterns of lunar phases, eclipses of the sun and moon, and seasons. [*Clarification Statement:* Examples of models may be physical, graphical, or conceptual.]
Associated SEP: Developing and Using Models: Develop and use a model to describe phenomena.
Associated DCI: Patterns of the apparent motion of the sun, the moon, and stars in the sky can be observed, describes, predicted, and explained with models.
Associated CC: Patterns: Patterns can be used to identify cause and effect.

Figure 2.5. Performance expectation from middle school Earth and space sciences. Adapted from *Next Generation Science Standards, For States, By States, Volume 1, The Standards* (p. 78), by NGSS Lead States, 2013a, Washington, DC: The National Academies Press. Copyright 2013 by The National Academies Press. Adapted with permission.

advanced students. Therefore, teachers need to become familiar with the standards at their own level and across levels so that they know when to extend and accelerate each student's knowledge and skills.

Talent Trajectory: Differentiating the Pathway for Gifted and Talented STEM Students

As noted in *Preparing the Next Generation of STEM Innovators*, "The long-term prosperity of our Nation will increasingly rely on talented and motivated individuals who will comprise the vanguard of scientific and technological innovations; every student in America deserves the opportunity to achieve his or her full potential" (NSB, 2010, p. v). Educators need to increase the number and levels of these promising science students and not limit the numbers of students in advanced science and specialized STEM programs. Whether students plan to enter a STEM field in a career anywhere from astronaut to zoologist or to simply become well-informed citizens who can make sense of the world, recognize patterns, make generalizations and test conjectures, make and defend logical decisions, and critique the reasoning of others, science is critical to their development.

Preparation of high-level STEM students should not be rushed. Appropriate pacing for our top students should include not only acceleration, but also time for our students to experience the joy of investigating rich concepts in depth and applying innovative scientific reasoning and justification to a variety of scientific, mathematical, engineering, and other problems. They should have engaging, problem-based learning that encourages students to grapple with scientific challenges to deepen their understanding of complex concepts (NAGC, 2008). There should be a seamless articulation from elementary to middle to high school where courses are carefully planned and there is no repetition of courses or content within the courses, especially as students move from elementary to middle school or from middle school to high school. This should include flexibility in scheduling, location, seat time, and other potential impediments to ensure that students make continuous progress throughout the K–12 programs in those areas that hold great interest and appeal. For example, students who are interested in science and preparing for STEM careers should have opportunities to take math and science every year of high school from highly motivated teachers. There should be widespread availability of specialized STEM schools, programs, or advanced classes for students from elementary through high school to fuel their passions and give them the preparation necessary to move ahead.

Following the first 3 years of high school physics, chemistry, and biology, a menu of science course options should be available. These options for advanced courses might include advanced, honors, or AP courses in biology, chemistry and physics, and specialized courses such as organic chemistry or microbiology. Courses should be designed to be challenging, engaging, and relevant, preparing and exciting students about their futures in college and careers. These courses may be provided at the high school, online, or through dual enrollment at a nearby college or university. Note that students who are the most highly advanced in STEM need mentorships, laboratory opportunities for individual research with professionals in their areas of interest, or more radical acceleration such as early entrance to college.

Appendix K of the NGSS (NGSS Lead States, 2013b) provides three different course maps that are models of different ways to map the performance expectations of grades 6–8 and 9–12 onto courses: the conceptual progressions model, the science domain model, and the modified science domain model. There is a discussion of benefits and challenges to each; however, these are course maps for students on a typical trajectory, not a talent trajectory. There are no suggestions for talent trajectories for advanced students in the NGSS or the *Framework for K–12 Science Education*, but as in mathematics, advanced and talented science students need access to advanced classes earlier and more often than typical learners.

Purposeful planning should occur for those students who decide after middle school that they are interested in taking advanced science such as AP and International Baccalaureate (IB) before leaving high school. Additional opportunities for acceleration should occur. For example:

- allowing students to take two science courses simultaneously;
- allowing students in schools with block scheduling to take a science course in both semesters of the same academic year;
- offering summer courses that are designed to provide the equivalent experience of a full-year course in all regards;
- creating different compaction ratios, including 4 years of high school content into 3 years beginning in ninth grade;
- creating hybrid courses; and
- allowing students to participate in programs such as the AP Cambridge Capstone Program (learn more at http://aphighered.collegeboard.org/exams/cambridge–capstone).

These recommendations are more in line with the acceleration research literature and position statements in gifted education that suggest intellectually talented youth achieve at an impressively high level if they receive an appropriately challenging education (Benbow & Stanley, 1996; Colangelo et al., 2004; Swiatek & Benbow, 1991, 1992). The National Association for Gifted Children's (2004) position statement described "grade skipping, telescoping, early entrance into kindergarten or college, credit by examination, and acceleration in content areas" as research-based gifted education practices and concluded by stating, "Highly able students with capability and motivation to succeed in placements beyond traditional age/grade parameters should be provided the opportunity to enroll in appropriate classes and educational settings" (para. 7).

All students, including high-ability learners, should have access to pathways to excellence in science. Currently, many student opportunities exist in the form of out-of-school enrichment activities (i.e., science/math clubs, competitions, summer camps). To increase opportunities for talent development, acceleration, ample time, and advanced course matriculation must be considered (NSB, 2010). Creating multiple pathways that work in concert as synchronized education rather than as stand-alone options increases the likelihood for student success in STEM disciplines.

Differentiating the Learning Progressions

The learning progressions in the NGSS provide a framework or a general map for learning for typical learners, and thus the LPs will need to be differentiated for gifted and advanced science learners. Two familiar tools in gifted education that may be useful for differentiating the LPs are Tomlinson's (2001) Equalizer and the idea of Ascending Intellectual Demand (AID) from the Parallel Curriculum Model (Tomlinson et al, 2009). Both offer paradigms for thinking about how to move students from novice to expert knowledge, conceptual understanding, and practices.

For example, using the Equalizer to plan a differentiated lesson for L1-C, a teacher might choose to use the "concrete to abstract" continuum. Students who still need to understand how animals obtain food may work with task cards that ask them to match animals to the food they eat and think about how that animal finds the food. Students with more sophisticated thinking and a clear understanding of how animals get food as well as use the food for growth might need to work beyond the grade level and interact with a more abstract topic such as how plants acquire material for growth.

The Ascending Intellectual Demand continuum from novice to expert in science (Tomlinson et al., 2009) defines what novice, apprentice, practitioner, and expert learners know, understand, and are able to do at each level. See Table 2.5 for an example.

In addition, DCI PS2 Motion and Stability: Forces and Interactions at the middle school level (grades 6–8) includes the following performance expectation (MS-PS2-5): "Conduct an investigation and evaluate the experimental design to provide evidence that fields exist between objects exerting forces on each other even though the objects are not in contact" (NGSS Lead States, 2013a, p. 59). Using the Practices LP in Table 2.4 and the descriptions of novice to expert performance in Table 2.5, teachers can apply their knowledge of each child's abilities in this area to differentiate the lesson by placing each one at the appropriate entry point to afford that child a learning experience with appropriate challenge and complexity.

In addition to these two examples, there are many other ways to differentiate for advanced learners. Some of these strategies include:

1. *Providing appropriate pacing and acceleration.* By studying the LPs, teachers can identify concepts that are above level. Including these above-level concepts on pre- and ongoing assessments raises the ceiling and allows the teacher to identify what students know. Teachers can then integrate new, challenging concepts into class problems or investigations or compact the curriculum, allowing

Table 2.4

Partial Practice 3: Planning and Carrying Out Investigations Progressions

Grades K–2	Grades 3–5	Grades 6–8	Grades 9–12
Plan and conduct an investigation collaboratively to produce data to serve as the basis for evidence to answer a questions.	Plan and conduct an investigation collaboratively to produce data to serve as the basis for evidence, using fair tests in which variables are controlled and the number of trials is considered.	Plan and conduct an investigation individually and collaboratively, and in the design identify independent and dependent variables and controls, what tools are needed to do the gathering, how measurements will be recorded, and how many data are needed to support a claim.	Plan and investigation or test a design individually or collaboratively to produce data to serve as the basis for evidence as part of building and revising models, supporting explanations for phenomena or testing solutions to problems. Consider possible confounding variables or effects and evaluate the investigation's design to ensure variable are controlled.

Note: Adapted from *The Next Generation Science Standards: Appendix F* (pp. 7–8), by NGSS Lead States, 2013d, Washington, DC: The National Academies Press. Copyright 2013 by The National Academies Press. Adapted with permission.

students to do alternative investigations that match their current level of achievement. For example, some students might be designing their own experiments manipulating multiple variables instead of just one.

2. *Integrating greater complexity and depth in problems.* Teachers can develop lessons that include more abstract concepts or that involve multiple connections with other scientific ideas.

3. *Focusing on broader concepts.* Teachers might develop units around broader science concepts or themes. For example, students might examine the concept of systems by studying biological systems and relating what they learn about how systems operate to the solar system.

Table 2.5

Partial Novice to Expert Continuum in Science

Novice	Apprentice	Practitioner	Expert
Sees experimentation as an end in itself rather than a means to an end; Inadvertently includes and fails to manage multiple variables.	Manipulates one variable within an experiment with ease; Understands, identifies, and analyzes the relationships among the independent and dependent variables, constants, and controls.	Effectively manipulates multiple variables within an experiment; Plans for and observes a wide range of factors (variables, constants, and controls) and discerns patterns.	Conducts complex experiments with ease and fluidity; Freely manipulates methods, tools, knowledge, and self to achieve desired results.

Note: Adapted from *The Parallel Curriculum* (p. 247), by C. A. Tomlinson, S. N. Kaplan, J. S. Renzulli, J. H. Purcell, J. H. Leppien, D. E. Burns, C. A. Strickland, & M. B. Imbeau, 2009, Thousand Oaks, CA: Corwin. Copyright 2009 by Corwin. Adapted with permission.

4. *Incorporating more creativity in science.* Teachers might develop more ambiguous engineering problems where the student needs to create a new tool to solve the problem. Here, creativity plays out in collaborative brainstorming and design testing.

5. *Asking higher level thinking questions.* Teachers should ask questions that encourage students to make comparisons, predictions, and draw conclusions about the problems and questions that they are identifying or investigating. For example, using the data that were collected on students performing a task to test the students' reflexes, what might you predict about the students' performance? Were your predictions correct? Draw conclusions about the students' performance and describe the trend. What factors might have influenced the students' performance? Moreover, students should have opportunities to reflect on ways that they solved problems and about their own reasoning—metacognitive thinking. How did you arrive at your solution? Are there other ways you might have solved the problem?

6. *Grouping students with similar interests and abilities.* Teachers can use formative assessments to identify those students who are ready for above-level content and homogeneously group these students together. Students might also be grouped around areas of interest. For example, several students might want to design a study to examine the behavior of phototropic plants to varying amounts of light.

7. *Identifying collaborators.* Teachers need to identify scientists and engineers at the university or within the community to provide not only expert instruction and opportunities to allow students to work with authentic, complex problems and investigations, but also to provide an introduction to professional networks and the work habits that are required in a professional field. Collaborators might also be accessed using distance learning and might include other educators as well as students.

Getting Started in Differentiating the Learning Progressions

In getting started, curriculum directors and teachers in role-alike groups (e.g., K–2 teachers, 3–5 teachers, etc.) will want to examine the NGSS at their grade levels or grade bands. This process will allow teachers to identify learning progressions and the important concepts, processes, and student outcomes within and across domains and dimensions.

Once this initial work is completed, teachers across all grades will want to meet to compare learning progressions to ensure a comprehensive and coherent structure across all of the grades. At this time, acceleration options need to be discussed. These might include the different secondary options as well as ways to advance students through a learning progression as they demonstrate mastery of foundational concepts. Teachers will need to consider students who are above level. How will subsequent teachers provide for these gifted and advanced students? How will the curriculum be adapted? Will there be administrative options such as single-subject acceleration, grade skipping, or early entrance to high school or college?

After this work is completed, the teachers will want to examine the curriculum and the units that they have already developed, review student work samples, and identify how these relate to the standards' learning progressions. With many topics in science being eliminated or moved to other grade levels, this will be an important step. This process will uncover whether or not the curricular units build the student's understanding of key concepts over time. The favorite dinosaur unit may no longer be applicable at its current grade level. Teachers will also want to look at ways that they can adapt or differentiate the units for gifted learners (see pages 47–90 for sample differentiated learning experiences for typical and advanced learners).

Once the curricular units are designed and differentiated, teacher teams will want to identify available human and material resources. At this point, they may even want to consider developing partnerships with universities, identifying competitions, and involving students in distance learning opportunities for extensions.

The final step in this process is to "test" the validity of the learning progression and its effects on all of the students. In some cases, learning progressions have already been validated (Consortium for Policy Research in Education, 2009). When a LP has not been previously validated, the teacher will want to use and develop formative, ongoing, and summative types of assessments to aid in mapping the progression for various learners.

Learning progressions, which differ from more traditional scopes and sequences, help teachers identify what is expected from students and key benchmarks within and across grade levels. Knowing how LPs are connected to one another, both vertically and horizontally, helps educators establish goals that challenge each learner, build a foundation for subsequent knowledge and skills, and develop above-level assessments. Because the LPs were

developed for the typical learner, teachers will need to differentiate them for gifted and advanced learners in science.

Chapter 3

Differentiated Learning Experiences

The following pages offer examples of activities to support the implementation of the NGSS. The sample activities are provided for elementary, middle, and high school standards, and were designed to give exemplars with one example from each domain (physical science, life science, Earth and space sciences, and engineering design) per level. Moreover, the exemplars include sections that denote Crosscutting Concepts, Disciplinary Core Ideas, and Science and Engineering Practices.

Creating Examples in Science for Advanced Learners

The examples that follow address important standards in science that need to be differentiated for gifted and advanced learners at key stages of development. The processes to accomplish that task are noted, along with ideas for implementation. In a book on differentiating the standards (Adams, Cotabish & Ricci, 2014), we focused on explicating standards by strand and across grade-level clusters, providing exemplars for differentiating the science standards. This text focuses on integrating engineering

 DOI: 10.4324/9781003238522-3

practices into learning experiences and also addresses differentiating learning experiences when using problem- and project-based activities. Furthermore, we have provided prototypical learning experiences that use the Standards for Engineering Practice, including the one that relates to developing innovative and creative students and that support the implementation of the NGSS. The examples provided in the next section of this book share some common features with respect to how they are represented as well as how the task demands were conceptualized. Each of the learning experiences includes these seven components:

1. NGSS are selected for elementary, middle, and high school across all four science domains. In this way the teacher is able to see both the scope and the sequence within one learning progression.

2. A brief description of the learning experience is provided for both typical and advanced students. This overview compares and contrasts the learning experience for typical and advanced students. Note that sometimes the initial problem is the same for both typical and advanced learners, using questions and formative assessments to differentiate and develop scientific creativity and giftedness.

3. A sequence of activities is outlined for each learning experience.

4. Implementation strategies are included at the end of each learning experience. These strategies may include preassessments, grouping, independent research, use of external data, interdisciplinary problems, and other ways of accelerating and enriching the learning experience.

5. Each learning experience includes suggestions for assessments. For the most part, these assessments provide questions and specific characteristics that might be used in the design of product and performance rubrics or other types of summative assessments. Formative assessment in these activities includes the use of pretests, differentiation of tasks, and questions to assess during the problem-

solving process, observation and analysis of student work, and authentic cross-disciplinary tasks and research.

6. A Foundation Box is provided that includes the DCI, CC, and SEP that are covered in the lesson.

7. The Connections Box provides linkages to the CCSS for both math and ELA that are covered in the LE.

Additionally, the experiences include two elements embedded within the NGSS associated with the performance expectations that are meant to render additional support and clarity:

- *Assessment boundary statements* are included with individual performance expectations where appropriate to provide further guidance or to specify the scope of the expectation at a particular grade level.
- *Clarification statements* are designed to supply examples or additional clarification to the performance expectations.

Assessment boundary and clarification statements should not limit teaching and learning. They are meant to provide guidance as educators plan instruction.

Sample Learning Experiences

Teacher Information: Typical learners will make observations and measurements of an object's motion to provide evidence that a pattern can be used to predict future motion. The advanced learning experiences will provide an effective introduction to accelerated motion if students are to understand that recorded motion is not uniform. Teachers will need to encourage students to find unique races for timed trials. Calculated speeds are average speeds, although the longer the distance traveled, the closer the average speed will approach the speed at the finish line if the speed is uniform.

Grade 3 3-PS2 DCI: Motion and Stability: Forces and Interaction	Typical Learner(s)	Advanced Learner(s)
	Essential Question: To what extent do patterns predict an object's motion?	
Performance Expectations: **3-PS2-2. Make observations and/or measurements of an object's motion to provide evidence that a pattern can be used to predict future motion.** *Clarification Statement:* Examples of motion with a predictable pattern could include a child swinging in a swing, a ball rolling back and forth in a bowl, and two children on a seesaw. *Assessment Boundary:* Assessment does not include technical terms such as period and frequency.	**Directions:** After introducing the term *motion* as an action or process of moving or being moved, have students work in groups of three or four to observe objects with predictable motion. Students will observe events producing predictable patterns such as rolling a ball back and forth in a bowl, sliding down a slide, a skateboard going down a ramp, a child swinging in a swing, and two children on a seesaw. Students are then asked to measure these events using a measuring stick and a stopwatch. Although the assessment boundary does not include the introduction of technical terms such as frequency and period, the teacher needs to define the period for students (e.g., a child swinging back and forth one time would be defined as 1). Have an assigned group recorder record all measurements. *continued*	**Directions:** After advanced learners have demonstrated mastery of the concepts presented to typical learners, advanced learners will observe activities that allow them to discover that recorded motion is not uniform and predictable. Have students work in groups of three or four and determine the speed of selected events. Students could invent certain races such as hopping on the same foot, rolling on the lawn, backward walk, or three-legged race. Have an assigned group recorder record all measurements. These measurements will be used to make the necessary calculations. Have students always show the units used in measured and calculated values at the top of each column along with the name of the quantity. Have students construct a table like the one below. *continued*

Grade 3: Motion and Stability, *continued*

These measurements will be used to capture patterns of motion and assist with predicting future motion. Students are to conduct five trials to obtain data for prediction purposes. Have students always show the units used in measured and calculated values in the top of each column along with the name of the quantity. Have students construct a table like the one below.

Example:

Activity	Distance	Time
Rolling Ball (in a bowl)		
Seesaw		
Child in a Swing		

To assess student understanding, ask students the following questions after conducting five trials: (a) What do you notice about the data collected? (b) What patterns emerged over time? (c) What do you predict will be the next value (in trial 6)? (d) Do you think motion is always predictable? Explain. (e) How do you think speed, distance, and time are related? (f) What was the fastest event in class?

For further challenge, the teacher would ask students to invent their own events to test the predictability of motion.

Example:

Activity	Distance	Time
Toy Car Race		
Walking Backwards		
Running		

As students are exploring these concepts, ask advanced learners to share possible equations and calculations to represent the object's motion (do not give students the equation). Discuss the patterns presented in the data. Once students express a conceptual understanding of the speed of something in a given direction, the teacher can introduce the term velocity.

To assess for student understanding, ask students the following questions: (a) What do you notice about the data collected? (b) What patterns emerged? (d) Do you think motion is predictable? Explain. (e) How do you think speed, distance, and time are related? (f) What was the fastest event in class? (e) Will the fastest event always be the fastest event? Why or why not?

For further challenge, advanced students could create their own events to test the motion and calculate velocity. The teacher could ask students to predict how direction and friction may affect motion and velocity. Students could test the effects of each and gather additional data.
continued

Grade 3: Motion and Stability, *continued*

	Sample observations/calculations include: v = average speed s = distance in meters t = time in seconds
Implementation	**Materials:** measuring sticks, stopwatches, or motion detector. Have students list their events and average velocities on the board. Do not expect the same types of results from each student. As students are exploring the concepts of velocity (average velocity = distance divided by time), do not give students the equation. Encourage students to try to figure out on their own how to use the data they collect (e.g., some students will try to divide the time by distance). This is acceptable because class discussion will bring about the understanding of how to use the data.
Foundation Box	**Science and Engineering Practices:** • Develop and use a model to describe phenomena. **Disciplinary Core Ideas:** • **PS2.A: Forces and Motion:** The patterns of an object's motion in various situations can be observed and measured; when that past motion exhibits a regular pattern, future motion can be predicted from it. (Assessment Boundary: Technical terms, such as magnitude, velocity, momentum, and vector quantity, are not introduced at this level, but the concept that some quantities need both size and direction to be described is developed.) **Crosscutting Concepts:** • **Patterns:** Patterns of change can be used to make predictions.
Connections to Common Core State Standards	**ELA/Literacy:** **W.3.7:** Conduct short research projects that build knowledge about a topic. **W.3.8:** Recall information from experiences or gather information from print and digital sources; take brief notes on sources and sort evidence into provided categories.

Note: Activity for advanced learners is adapted from PRISMS Plus (CD version) by T. M. Cooney, L. T. Escalada, and R. D. Unruh, 2008, Cedar Falls: University of Northern Iowa Physics Department. Copyright 2008 by UNI Physics Department. Adapted with permission.

Grade 3 3-LS4 DCI: Biological Evolution: Unity and Diversity	**Teacher Information:** Students will understand why animals need to adapt to their environments. Students will see how coloring, markings, and physical actions can make an animal better adapted to its environment. Typical learners will explore the reasons why camouflage is important to animals of all kinds and predict the environment an animal would live in based on their knowledge of cloaking (camouflage). Advanced learners will create a fictitious organism/animal from evidence of common ancestry (e.g., fossil evidence) that utilizes camouflage for survival, with special consideration of the animal's habitat and ecosystem.	
	Typical Learner(s)	**Advanced Learner(s)**
	Essential Question: How and why do animals adapt to their environments?	
Performance Expectation: 3-LS4-3. Construct an argument with evidence that in a particular habitat some organisms can survive well, some survive less well, and some cannot survive at all. *Clarification Statement:* Examples of evidence could include needs and characteristics of the organisms and habitats involved. The organisms and their habitat make up a system in which the parts depend on each other.	**Directions:** The teacher will facilitate a discussion about organism/animal survival techniques by relating to students' personal experiences. To begin, the teacher asks students how hunters use hunting apparel to hunt prey such as deer (leading to a discussion about camouflage; remind students that camouflage is a concept, not a color). The teacher will ask students to share their personal experiences with camouflage apparel. Be sure to have students clarify the difference between the terms "predator" and "prey." Explain to the students that they are going to be conducting an experiment that will illustrate the importance of camouflage to specific organisms/animals and their habitat. (The animals use cloaking/camouflage techniques for survival from predators and/or to catch prey.) *continued*	**Directions:** The teacher will facilitate a discussion about organism/animal survival techniques by relating to students' personal experiences. To begin, the teacher asks students how hunters use hunting apparel to hunt prey such as deer (leading to a discussion about camouflage; remind students that camouflage is a concept, not a color). The teacher will ask students to share their personal experiences with camouflage apparel. Be sure to have students clarify the difference between the terms "predator" and "prey." Explain to the students that they are going to be creating an animal that will illustrate the importance of camouflage to specific habitats/environments. (The animals use cloaking/camouflage techniques for survival from predators and/or to catch prey.) *continued*

Grade 3: Biological Evolution, *continued*

Probing Questions: "Let's take a look at the different ways animals use camouflage techniques. You will discover why certain animals use cloaking in specific environments."

The teacher will facilitate the following steps:

1. Have students cut out 12 butterflies from the patterned paper and 12 from each of the solid papers.

2. Have one student in the group place one full piece of patterned paper on the floor and place the 36 butterflies on it carefully.

3. Set the timer to 10 seconds with one student's eyes covered.

4. Start timer and have the student pick up as many butterflies as he or she can in the 10 seconds.

5. Compare results from the camouflaged butterflies to the solid ones.

6. Display pictures of various animals that adapt to their surroundings.

7. Discuss how protective coloring often helps animals hide from their predators.

8. Divide students into groups of two.

9. Give each group two pictures of camouflaged animals.

10. Ask students to circle the camouflaged animals in the pictures with a highlighter.

continued

Conducting a search from Internet resources, advanced learners (in small groups) will collect evidence from various kinds of animals that once lived on Earth and are no longer found anywhere. From the evidence, they will **create** a fictitious organism/animal that will utilize camouflage for survival, with special consideration given to the animal's habitat and ecosystem.

To **assess** student understanding, advanced learners will answer the following questions:

- What evidence supports your animal's use of camouflage as a survival tool?
- Why does your animal live in this environment?
- Is your animal a predator? Prey? Both? Explain.
- What role does this animal play in the ecosystem?
- Predict what would happen to your animal over time.
- Would this animal be able to survive in today's environment? Why or Why not?

Conclusion: Conclude the lesson by having groups share their findings with the class.

Grade 3: Biological Evolution, *continued*	
	11. Ask students to determine if the animal is a predator, prey, or both. Also, ask them to contemplate the animals' role in the ecosystem. To **assess** student understanding, ask students to the following questions: • What animal is camouflaged here? • Why do you think this animal lives in this environment? • Is this animal preying on another animal's prey? How do you know? • What role does this animal play in the ecosystem? **Conclusion:** Conclude the lesson by having groups share their findings with the class.
Implementation	**Materials:** Several sheets of wrapping paper of three different kinds (one patterned and the other two different solid colors), scissors, timer, pictures of habitats and animals, and highlighters. Contact or shelving paper with patterns works best for this activity. Have students practice science safety and put on their goggles. It is recommended that the teacher group advanced learners to conduct their investigation/simulation. If time permits, the teacher could integrate the timer into advanced learning experience (most students deem it as fun).
Foundation Box	**Science and Engineering Practices:** • Constructing Explanations and Designing Solutions **Disciplinary Core Ideas:** • LS4.C: Adaptation: For any particular environment, some kinds of organisms survive well, some survive less well, and some cannot survive at all. <div align="right">*continued*</div>

Grade 3: Biological Evolution, *continued*	
	Crosscutting Concepts: • **Cause and Effect:** Cause and effect relationships are routinely identified and used to explain change.
Connections to Common Core State Standards	**ELA/Literacy:** **RI.3.1:** Ask and answer questions to demonstrate understanding of a text, referring explicitly to the text as the basis for the answers. 　**RI.3.2:** Determine the main idea of a text; recount the key details and explain how they support the main idea. 　**RI.3.3:** Describe the relationship between a series of historical events, scientific ideas or concepts, or steps in technical procedures in a text, using language that pertains to time, sequence, and cause/effect. 　**W.3.1:** Write opinion pieces on topics or texts, supporting a point of view with reasons. 　**W.3.2:** Write informative/explanatory texts to examine a topic and convey ideas and information clearly. 　**SL.3.4:** Report on a topic or text, tell a story, or recount an experience with appropriate facts and relevant, descriptive details, speaking clearly at an understandable pace. **Mathematics:** **MP.2:** Reason abstractly and quantitatively. **3.MD.B.3** Draw a scaled picture graph and a scaled bar graph to represent a data set with several categories. Solve one- and two-step "how many more" and "how many less" problems using information presented in scaled bar graphs.

Note: Learning experience adapted from *The Wildlife Ecologist* by L. M. Williams, M. C. Brittingham, and S. S. Smith, 2001, University Park: Pennsylvania State University. Copyright 2001 by Pennsylvania State University. Adapted with permission.

	Typical Learner(s)	Advanced Learner(s)
Grade 4 **4-ESS2** **DCI: Earth's Systems**	**Teacher Information:** Typical learners will observe the effects of weathering and erosion, identify the three agents of erosion, understand what erosion is, and identify preventative measures. Advanced learners will research and demonstrate a physical process that affects the availability and quality of natural resources. Furthermore, advanced learners will describe how the interaction has affected Earth overtime and predict future effects of the interaction.	
	Essential Questions: What effects does erosion have on the Earth?	
Performance Expectation: 4-ESS2-1. **Make observations and/or measurements to provide evidence of the effects of weathering or the rate of erosion by water, ice, wind, or vegetation.** *Clarification Statement:* Examples of variables to test could include the angle of slope in the downhill movement of water, amount of vegetation, speed of wind, relative rate of deposition, cycles of freezing and thawing of water, cycles of heating and cooling, and volume of water flow. *Assessment Boundary:* Assessment is limited to a single form of weathering or erosion.	**Directions:** To introduce students to the concept of erosion, the teacher will show the following YouTube video: http://www.youtube.com/watch?v=G5Rp9MJJGCU. After watching the video, the teacher will ask the following probing questions: • What is erosion? • What are three components of erosion? • What are some good examples of erosion? After addressing students' misconceptions, the teacher will facilitate three erosion lab activities using the listed materials. Students will be divided into five groups and will rotate through each lab activity until students have rotated through to complete all five activities. **Station 1:** Using a tray filled with sand on one end, students will pour water on the other end and slide the *continued*	**Directions:** To introduce students to the concept of erosion, the teacher will show the following YouTube video: http://www.youtube.com/watch?v=G5Rp9MJJGCU. After watching the video, the teacher will ask the following probing questions: • What is erosion? • What are three components of erosion? • What are some good examples of erosion? Using digital resources, advanced learners will research how physical processes (e.g., climate, plate tectonics, erosion, soil formation, water cycle, and circulation patterns in the ocean shape patterns) in the environment influence the availability and quality of natural resources. Given the materials listed in the *materials section*, advanced learners will **create** a simulation of erosion *continued*

Grade 4: Earth's Systems, *continued*

pan back and forth to create a wave. Record observations.

Station 2: Using a tray filled with sand, students will blow on the sand with a hairdryer to simulate wind. Varying wind speeds can be simulated with the multiple blowing speeds of the hair dryer. Record observations.

Station 3: Using an inclined tray filled with soil, students will pour varying amounts of water (varying the volume of water flow; record the amount of water variance) on the soil. The trial will be repeated with the inclusion of rocks and varying the angle of the slope (inclined tray) to observe the downhill movement of water and rate of erosion. Record observations.

Station 4: Using clumps of soil, rocks, one tray, a pitcher of water, and plants, students will design at least three trials to determine the best course of action to prevent erosion.

Station 5: Place an ice cube in a plastic cup of warm water. Record observations.

To **assess** for student understanding, the teacher will conclude with the following questions:

- How are the labs similar to the processes of erosion?
- In what ways do rivers or streams act in the same way?
- What will happen if erosión continues?

continued

that demonstrates one of these physical processes in action. Students are expected to represent measurement quantities using diagrams that feature a measurement scale.

To **assess** for student understanding, the teacher will conclude with the following questions:

- How is your simulation similar to the processes of erosion?
- In what ways does your simulation act in the same way?
- In accordance with your simulation, what do you predict will happen if erosion continues?
- What do you predict will happen when there are cycles of freezing or thawing water? Name some examples.
- What are some preventative measures one could take to prevent erosion in your simulated activities?

The teacher will ask students to reflect upon their learning (e.g., journal writing) and consider how similar interactions have affected Earth over time. For additional complexity, student can predict future effects of other physical processes.

Grade 4: Earth's Systems, *continued*

	• What do you predict will happen when there are cycles of freezing or thawing water? Name some examples. • What are some preventative measures you took to prevent erosion in your simulated activities? • How do plants help in erosion prevention? After facilitating the discussion, the teacher will ask students to reflect upon their learning (e.g., journal writing) and consider additional examples of erosion that have affected the topography of Earth over time. • How does weathering and erosion occur? • What is the result of each type of weathering? • Where on Earth do these various types of erosion occur?
Implementation	**Materials:** Tub of loose, unconsolidated dirt, rough rocks, two pitchers, water, cups, inclined tray, two trays, potting plants (six pack), ice cubes, plastic cup, and hairdryer. Access to Internet is needed to carry out the tasks assigned to advanced learners. Have students practice science safety and put on their goggles. With regard to advanced learners, the teacher may want to specify the measurement scale when assigning the diagram component of the learning experience. As written, this component is intentionally left open ended to promote creative ideas.
Foundation Box	**Science and Engineering Practices:** • Planning and carrying out investigations

continued

Grade 4: Earth's Systems, *continued*	
	Disciplinary Core Ideas: • ESS2.A: Earth Materials and Systems: Rainfall helps to shape the land and affects the types of living things found in a region. Water, ice, wind, living organisms, and gravity break rocks, soils, and sediments into smaller particles and move them around. • ESS2.E: Biogeology: Living things affect the physical characteristics of their regions. **Crosscutting Concepts:** • Cause and Effect: Cause and effect relationships are routinely identified, tested, and used to explain change.

Connections to Common Core State Standards	**ELA/Literacy:**
	    W.4.8: Recall relevant information from experiences or gather relevant information from print and digital sources; take notes and categorize information, and provide a list of sources.
	
	Mathematics:
	    MP.2: Reason abstractly and quantitatively.
	    MP.4: Model with mathematics.
	    MP.5: Use appropriate tools strategically.
	    4.MD.A.1: Know relative sizes of measurement units within one system of units including km, m, cm; kg, g; lb, oz.; l, ml; hr, min, sec. Within a single system of measurement, express measurements in a larger unit in terms of a smaller unit. Record measurement equivalents in a two-column table.
	    4.MD.A.2 Use the four operations to solve word problems involving distances, intervals of time, liquid volumes, masses of objects, and money, including problems involving simple fractions or decimals, and problems that require expressing measurements given in a larger unit in terms of a smaller unit. Represent measurement quantities using diagrams such as number line diagrams that feature a measurement scale.

Note: Learning experience is adapted from *Erosion* by J. Haley and L. Monk, 2013, Conway: University of Central Arkansas. Copyright 2013 by University of Central Arkansas. Adapted with permission

Grade 1 1-LS1 K–2 Engineering Design	**Teacher Information:** Students will demonstrate (through a design challenge) their understanding of the function of plant and animal external parts by using them to solve human problems. Advanced learners will conduct the same activity; however, complexity is added to the design challenge.	
	Typical Learner(s)	**Advanced Learner(s)**
	Essential Questions: How do plants and/or animals use their external parts to help them survive, grow, and meet their needs?	
Performance Expectation: **K–2-ETS1-1. Ask questions, make observations, and gather information about a situation people want to change to define a simple problem that can be solved through the development of a new or improved object or tool.** **K–2-ETS1-2. Develop a simple sketch, drawing, or physical model to illustrate how the shape of an object helps it function as needed to solve a given problem.** **K–2-ETS1-3. Analyze data from tests of two objects designed to solve the same problem to compare the strengths and weaknesses of how each performs.** *continued*	**Background Information:** Plants and animals have parts that help them survive in their environment and protect them from predators. For example, a turtle has a shell that provides protection from predators, skunks release a bad smell when they are scared, and a cactus has needles that protect it from animals and help it retain water. Through a water balloon design challenge, students are to mimic how plants and animals use their external parts to help them survive. **Design Challenge:** Skateboarding is one of your favorite hobbies. You particularly enjoy doing tricks off of high skateboarding ramps. However, you keep falling off your skateboard hurting your arms. As you begin thinking about how other animals use their parts to protect themselves, you decide to conduct an experiment using a water balloon activity to simulate/mimic how a shape of an object helps it function as needed for protection. Using at least four *continued*	**Background Information:** Plants and animals have parts that help them survive in their environment and protect them from predators. For example, a turtle has a shell that provides protection from predators, skunks release a bad smell when they are scared, and a cactus has needles that protect it from animals and help it retain water. Through a water balloon design challenge, students are to mimic how plants and animals use their external parts to help them survive. **Design Challenge:** Skateboarding is one of your favorite hobbies. You particularly enjoy doing tricks off of high skateboarding ramps. However, you keep falling off your skateboard hurting your arms. As you begin thinking about how other animals use their parts to protect themselves, you decide to conduct an experiment using a water balloon activity to simulate/mimic how a shape of an object helps it function as needed for protection. Using only three *continued*

Grade 1: Engineering Design, *continued*		
1-LS1-1. Use materials to design a solution to a human problem by mimicking how plants and/or animals use their external parts to help them survive, grow, and meet their needs. *Clarification Statement:* Examples of human problems that can be solved by mimicking plant or animal solutions could include designing clothing or equipment to protect bicyclists by mimicking turtle shells, acorn shells, and animal scales; stabilizing structures by mimicking animal tails and roots on plants; keeping out intruders by mimicking thorns on branches and animal quills; and detecting intruders by mimicking eyes and ears.	materials (see materials list for details), design a product that will protect your water balloons when dropped from various heights. **Criteria:** • Students must use *at least four parts* from the materials list. • Students will design and build a container that will protect a water balloon from breaking when dropped from various heights. **Students Will:** 1. Define the problem. 2. Brainstorm their design (through illustration). 3. Create a solution (by keeping notes about the problems they have and how they intend to solve them). 4. Test their solution. 5. Students will answer the following question: *Does your water balloon break when you drop it from 2 feet? 3 feet? 4 feet?* 6. Evaluate their solution. To **assess** for student understanding, the students will answer the following questions: *continued*	materials (see materials list for details), design a product that will protect your water balloons when dropped from various heights. **Criteria:** • Students must *only use three parts* from the materials lists. • Students will design and build a container that will protect a water balloon from breaking when dropped from various heights. *Note:* To add complexity, the teacher may choose to limit the weight of each package to a small weight beyond that of the water-filled balloon. This requires an accurate scale, and may require some testing on the part of the teacher to determine what the maximum allowable weight should be. **Students Will:** 1. Define the problem. 2. Brainstorm their design (through illustration). 3. Create a solution (by keeping notes about the problems they have and how they intend to solve them). 4. Test their solution. *continued*

Grade 1: Engineering Design, *continued*

• Was it the best solution? • Would one of your other ideas have been better? Why or why not? • What would you have done differently? • Could you add to it to make it better? What would you add to it? • How does the water balloon activity mimic how plants and/or animals use their external parts to help them survive? Give some examples. To conclude, students will share their results with the class.	5. Students will answer the following question: *Does your water balloon break when you drop it from 4 feet? 5 feet? 6 feet?* 6. Evaluate their solution. To **assess** for student understanding, the students will answer the following questions: • Was it the best solution? • Would one of your other ideas have been better? Why or why not? • What would you have done differently? • Could you add to it to make it better? What would you add to it? • Predict what would happen when the water balloon is dropped at 7 and 8 feet. Explain your prediction (rationale). • How does the water balloon activity mimic how plants and/or animals use their external parts to help them survive? Give some examples. In collaboration with peers and guidance and support from the teacher, advanced learners can use digital tools to produce a visual display. To conclude, students will share their results with the class.

Grade 1: Engineering Design, *continued*	
Implementation	**Materials:** water balloons, water, shoebox, Styrofoam half-balls and squares, feathers, paper plates, cardboard tubes, glue, tape, plastic tarp (can be used to cut into smaller pieces), and string/yarn. In lieu of using water balloons, the teacher could substitute raw eggs (making it an egg drop activity). **Assessment note:** When assessing for student understanding of engineering design, consider the following: • How well did students design their devices for different drops? • Do they understand what worked and what didn't? • Do they understand why particular designs didn't work? • Do they make the connection between the activity (designing a solution to protect the water balloon) and mimicking how plants and/or animals use their external parts to help them survive? **Additional Consideration for Assessing Advanced Learners:** When assessing for gifted student understanding of engineering design, also consider the following: • Can they make reasonable predictions about the fate of their water balloon when dropped at greater heights? • Can they explain their rationale behind their predictions?
Foundation Box	**Science and Engineering Practices:** • Asking Questions and Defining Problems • Developing and Using Models • Analyzing and Interpreting Data **Disciplinary Core Ideas:** • ETS1.A: Defining and Delimiting Engineering Problems • ETS1.B: Developing Possible Solutions • ETS1.C: Optimizing the Design Solution **Crosscutting Concepts:** • Structure and Function: The shape and stability of structures of natural and designed objects are related to their function(s).

Grade 1: Engineering Design, *continued*	
Connections to Common Core State Standards	**ELA/Literacy:** **RI.2.1:** Ask and answer such questions as who, what, where, when, why, and how to demonstrate understanding of key details in a text. **W.2.6:** With guidance and support from adults, use a variety of digital tools to produce and publish writing, including in collaboration with peers. **W.2.8:** Recall information from experiences or gather information from provided sources to answer a question. **SL.2.5:** Create audio recordings of stories or poems; add drawings or other visual displays to stories or recounts of experiences when appropriate to clarify ideas, thoughts, and feelings. **Mathematics:** **MP.2:** Reason abstractly and quantitatively. **MP.4:** Model with mathematics. **MP.5:** Use appropriate tools strategically. **2.MD.D.10:** Draw a picture graph and a bar graph (with single-unit scale) to represent a data set with up to four categories. Solve simple put-together, take-apart, and compare problems using information presented in a bar graph.

Note: Lesson implementation is adapted from *What Can We Learn from Plants and Animals?* by D. Dailey, 2014. Unpublished lesson. Adapted with permission.

Middle School
MS-PS2
DCI: Motion and Stability: Forces and Interactions

Teacher Information: This lesson focuses on Newton's Third Law of Motion. Before engaging in the learning experience, it is important that students already have a clear understanding of Newton's First and Second Laws of Motion (they may have knowledge of the concept, but not the names at this point) and have had a prior lesson introducing Newton's Third Law of Motion. As a pretest, take students outside or to the gymnasium to play tug-of-war. Have students come back to the classroom and in their science journals explain how the game is set up, played, and the end result using Newton's Second and Third Laws of Motion. Have them show all the forces involved in the game and how those relate to the Earth system. Those students who showed a deep understanding of Newton's Third Law (all forces on drawing are correct, explanation is detailed, accurate, and shows clear grasp of the concept) should complete the Advanced Learner activity. Others should complete the Typical Learner activity. Both learning experiences use balloon rockets to demonstrate Newton's Third Law, but the Advanced Learners must figure out how to get the balloon rocket down the guide wire *and* back.

Typical Learner(s)	Advanced Learner(s)
Essential Question: What happens when one object exerts a force on another object and the second object exerts an equal and opposite force back on the first one?	
Directions: Place students into groups of two or three and explain to them where the materials are located and what they need (or allow them to choose). Using only the materials that you have put out for them, direct students to make a balloon rocket system that will propel itself across the classroom on the wire or string provided. Caution them not to walk into the stretched guide wires.	**Directions:** Place students into groups of two or three and explain to them where the materials are located and what they need (or allow them to choose). Using only the materials that you have put out for them, direct students to make a balloon rocket system that will propel itself across the classroom on the wire or string provided and back again. Caution them not to walk into the stretched guide wires.
Allow time for exploration and construction of the balloon rockets. Students may need several trials before *continued*	Allow time for exploration and construction of the balloon rockets. Students may need several trials before *continued*

Performance Expectation: MS-PS2-1. Apply Newton's Third law to design a solution to a problem involving the motion of two colliding objects.

Clarification Statement: Examples of practical problems could include the impact of collisions between two cars, between a car and stationary objects, and between a meteor and a space vehicle.

continued

Middle School: Motion and Stability, *continued*

Assessment Boundary: Assessment is limited to vertical or horizontal interactions in one dimension.	the balloon rocket makes it to the end of the guide wire. Have them keep accurate data in their science notebooks to indicate if they changed their design. They should note and explain any differences in the results from one trial to another. They should measure distance traveled and the time it took for each trial. You may develop a data record sheet for those who have difficulty with this skill. Once all trials have been completed and data gathered, have students answer the following questions: • Why does your rocket move along the guide wire when the air escapes? Base your answer on your understanding of Newton's Second Law and Newton's Third Law. Draw a diagram of your system. On your diagram, show an arrow representing the force accelerating the rocket. What is the direction of this force? How do you know? • Which designs worked best? How might you modify your design if you had another chance to do the activity? • How do rockets work in outer space?	the balloon rocket makes it to the end of the guide wire and several trials to figure out how to get the balloon rocket to propel itself down and back. Have them keep accurate data in their science notebooks to indicate if they changed their design. They should also record distance and time per trial. They should note and explain any differences in the results from one trial to another. (*Note:* It is acceptable to manually release the second balloon [let students figure out they need two] to make the return trip possible.) Once all trials have been completed and data gathered, have students answer the following questions: • Why does your rocket move along the guide wire when the air escapes? Base your answer on your understanding of Newton's Second Law and Newton's Third Law. What was necessary to allow it to propel back? Draw a diagram of your system. On your diagram, show arrows representing the force accelerating the rocket in each direction. What is the direction of this force each time? How do you know? • Which designs worked best? How might you modify your design if you had another chance to do the activity? • How do rockets work in outer space?

Middle School: Motion and Stability, *continued*

Implementation	**Materials:** cylindrical balloons, masking tape, straws, paper clips, scissors, stopwatches, meter sticks, monofilament or string line(s) extending across the classroom to serve as guide wire for launching the balloon rockets. Monofilament fishing line works well as a guide wire. Feel free to use additional materials as desired. It is fun to put out other items for the students to ponder over. Items such as paper cups, sheets of paper, aluminum foil, packing material, and lightweight cardboard are good additions. Having cylindrical balloons of various sizes will also add challenge to the activity. Some teachers restrict each group to using the same materials; others do not use this restriction. Use whichever method seems appropriate to your particular situation. Have students practice science safety and put on their goggles.
	Prior to introducing the activity, string the guide wire in an appropriate area. You may choose to string it from a wall in the classroom to the wall directly opposite, or you may choose to go from corner to corner. Either way, it is a good idea to practice to be sure the balloons you have chosen will propel the distance you have selected. You may need to experiment several times to find a good match.
	Students should have their science notebook ready for note taking and record keeping. Questions and notebooks should be **assessed** for accuracy.
Foundation Box	**Science and Engineering Practices:** • Apply scientific ideas or principles to design an object, tool, or system. **Disciplinary Core Ideas:** • PS2.A: Forces and Motions **Crosscutting Concepts:** • **Systems and Models:** Models can be used to represent systems and their interactions—such as inputs, processes, and outputs—and energy and matter flow within systems. • **Influence of Science, Engineering and Technology on Society and the Natural World:** The uses of technologies and any limitations on their use are driven by individual or societal needs, desires, and values; by the findings of scientific research; and by differences in such factors as climate, natural resources, and economic conditions.

Middle School: Motion and Stability, *continued*

| Connections to Common Core State Standards | **ELA/Literacy:**
RST. 6–8.1: Cite specific textual evidence to support analysis of science and technical texts.
RST.6–8.3: Follow precisely a multi-step procedure when carrying out experiments, taking measurements, or performing technical tasks.
WHST.6–8.7: Conduct short research projects to answer a question (including a self-generated question), drawing on several sources and generating additional related, focused questions that allow for multiple avenues of exploration.

Mathematics:
MP.2: Reason abstractly and quantitatively
6.NS.C.5: Understanding that positive and negative numbers are used together to describe quantities having opposite directions or values; use positive and negative numbers to represent quantities in real-world contexts, explaining the meaning of zero in each situation.
6. EE.A.2: Write, read, and evaluate expressions in which letters stand for numbers.
7. EE.B.3: Solve multistep, real-world and mathematical problems posed with positive and negative rational numbers in any form, using tools strategically. Apply properties of operations to calculate with numbers in any form; convert between forms as appropriate; **assess** the reasonableness of numbers using mental computation and estimation strategies.
7.EE.B.4: Use variables to represent quantities in a real-world or mathematical problem, and construct simple equations and inequalities to solve problems by reasoning about the quantities. |

Note: This lesson implementation is adapted from PRISMS Plus (CD version) by T. M. Cooney, L.T. Escalada, and R.D. Unruh, 2008, Cedar Falls: IA, University of Northern Iowa Physics Department. Copyright 2008 by UNI Physics Department. Adapted with permission.

Middle School **MS-LS3** **DCI: Heredity: Inheritance and Variation of Traits**	**Teacher Information:** There is a wonderful website at Bryn Mawr University that has excellent activities geared to teaching science to middle and high school students. Some of the activities have been updated with connections to the NGSS. The web address is http://serendip.brynmawr.edu/sci_edu/waldron/. From there, you can select the particular topic in which you are interested. For the activity described below, we selected "Dragon Genetics I—Independent Assortment and Gene Linkage" for the typical activity, and "Dragon Genetics II—Understanding Inheritance" for the advanced learners. Before engaging in these learning experiences, students should understand basic principles of genetics, including relationships of genotype to phenotype, concepts of recessive and dominant alleles, and how understanding meiosis and fertilization provides the basis for understanding inheritance, as summarized in Punnett squares. Students who do not have a clear understanding of these concepts should begin with the activity "Genetics" found at the same site. The activity "Mitosis, Meiosis, and Fertilization" may be used as a preassessment for placing students in the appropriate activity. This activity may be accessed at http://serendip.brynmawr.edu/sci_edu/waldron/pdf/MitosisMeiosisTeachPrep.pdf. For each activity, there are Student Handout and Teacher Preparation Notes pages. You will need to look over the Teacher Preparation Notes several weeks before you plan to use these activities because in each activity there are materials that must be assembled.

Typical Learner(s)	**Advanced Learner(s)**
Essential Question: How do meiosis and fertilization result in transmission of genes from one generation to the next?	
Directions: Distribute the Student Handout so that each student has a handout. Students may work alone or in groups. Direct the students to complete each activity and answer all questions on the handout. Move around the room and make sure students understand what they are to do, notice who is unsure, who grasps the concepts easily, and who asks thoughtful questions. Check for confusion or misconceptions as you move from student *continued*	**Directions:** Although advanced learners will be focusing on independent assortment, segregation, and gene linkage as do typical learners, advanced learners will be using more traits per chromosome (nine as opposed to three). Moreover, they will work with a partner so that there is a "mother dragon" with chromosomes and alleles and a "father dragon" with chromosomes and alleles. They will work with autosomes, *continued*

Performance Expectation: MS-LS3-2: Develop and use a model to describe why asexual reproduction results in offspring with identical genetic information and sexual reproduction results in offspring with genetic variation.

continued

Middle School: Heredity, *continued*		
	Clarification Statement: Emphasis is on models such as Punnett squares, diagrams, and simulations to describe the cause and effect relationship of gene transmission from parent(s) to offspring and the resulting genetic variation.	to student and group to group. Distribute the materials when students are ready for that step. Some students may reach that point earlier than others. As an **assessment**, students' Punnett squares and dragon pictures should be checked for accuracy and a class discussion of the questions should ensue.	sex chromosomes, and incompletely dominant alleles. Distribute the Student Handout so that each student has a handout. Direct the students to complete each activity and answer all questions on the handout. Move around the room and make sure students understand what they are to do, notice who is unsure, who grasps the concepts easily, and who asks thoughtful questions. Check for confusion or misconceptions as you move from student to student and group to group. Distribute the materials when students are ready for that step. Some students may reach that point earlier than others. As an **assessment**, students' Punnett squares and dragon pictures should be checked for accuracy and a class discussion of the questions should ensue.
Implementation	For each activity, the appropriate materials must be prepared ahead of time. Please read the Teacher Preparation Notes as they provide explicit information for preparing and assembling the materials. For the Mitosis, Meiosis, and Fertilization activity (if you use it), you will need, hair rollers, masking tape, Velcro, and a permanent marker. For the other activities, you will need red, blue, yellow, and green craft sticks and permanent markers. In some cases, alternative choices for materials are suggested and all instructions are provided. The implementation is laid out in each set of Teacher Preparation Notes and the notes were followed as written. The Student Handout pages were used as written, but they may be downloaded as Word files and modified as necessary for your particular class. Each activity should take about one 50-minute class period.		

Middle School: Heredity, *continued*	
Foundation Box	**Science and Engineering Practices:** • Develop and use a model to describe phenomena **Disciplinary Core Ideas:** • LS1.B: Growth and Development of Organisms • LS3.A: Inheritance of Traits • LS3.B: Variation of Traits **Crosscutting Concepts:** • **Cause and Effect:** Cause and effect relationships may be used to predict phenomena in natural systems
Connections to Common Core State Standards	**ELA/Literacy:** **RST. 6–8.1:** Cite specific textual evidence to support analysis of science and technical texts. **RST.6–8.4:** Determine the meaning of symbols, key terms, and other domain-specific words and phrases as they are used in a specific scientific or technical context relevant to grades 6–8 topics. **RST.6–8.7:** Integrate quantitative or technical information expressed with words in a text with a version of that information expressed visually (e.g., in a flowchart, diagram, model, graph, or table). **SL.8.5:** Include multimedia components and visual displays in presentations to clarify claims and findings and emphasize salient points. **Mathematics:** **MP.4** Model with mathematics **6.SP.B.5** Summarize numerical data sets in relation to their context

Note: This lesson is adapted from Dr. Ingrid Waldron, University of Pennsylvania. Unpublished lesson. Adapted with permission.

Middle School MS-ESS2 DCI: Earth's Systems	**Teacher Information:** This lesson focuses on continental drift and plate tectonics. Typical students will work with materials to demonstrate the concepts by assembling Pangaea, pointing out fossils and rock formations that indicate the continents were once one large continent, and describing the movement of the continents to their current positions, and hypothesizing where the continents will be millions of years from now. There are a number of good activities that deal with assembling the continents into the super continent, Pangaea. One such activity is used here, but a search of the Internet will find quite a few others. Teachers should determine the appropriate level of challenge for their own typical students. Advanced students will look at seafloor spreading, another component to the theory of plate tectonics. They will calculate the rates of seafloor spreading from data on the magnetic properties and age of rocks from the ocean floor. Students who participate in this activity need to be able to calculate spreading rates in cm/year given distance and age (in millions of years). There are also variations on this activity, and these may be found by an Internet search. The student activities are from two sources, The International Ocean Discovery Program (IODP) and the U.S. Geological Survey. Good teacher background information can be found at http://www.slideshare.net/RobertCraig2/op-ch03-lectureearth3. Once all students have completed their respective activities, they should share their findings and discuss if, after hearing the evidence from those who completed other activities, they would change their original hypothesis. A good video for wrapping up the activities is found at http://www.youtube.com/watch?v=HrKTuCDierM.

Typical Learner(s)	**Advanced Learner(s)**
Essential Question: What evidence do we have that the Earth is always changing?	

| Performance Expectation:
MS-ESS2-3. Analyze and interpret data on the distribution of fossils and rocks, continental
continued | **Directions:** Students will complete an activity in which they will observe and analyze scientific evidence, use evidence to reconstruct Pangaea, interpret the evidence to form a hypothesis, and defend their hypotheses.
continued | **Directions:** Students will complete an activity about seafloor spreading and evidence from magnetic properties and ages of rocks in the ocean. Students will interpret a map of lithospheric plates, calculate seafloor
continued |

Middle School: Earth's Systems, *continued*		
shapes, and seafloor structures to provide evidence of past plate motions. *Clarification statement:* Examples of data include similarities of rock and fossil types on different continents, the shapes of the continents (including continental shelves), and the location of ocean structures (such as ridges, fracture zones, and trenches). *Assessment Boundary:* Paleomagnetic anomalies in oceanic and continental crusts are not assessed	The activity, *Wegener's Puzzling Evidence Exercise* (http://volcanoes.usgs.gov/about/edu/dynamicplanet/wegener/), allows students to observe some of the evidence Alfred Wegener used to propose his Theory of Continental Drift. Once they have completed the activity, the students will form their own opinions about whether the evidence they researched is compelling enough for them to accept the Theory of Continental Drift. Follow the steps below to complete the activity: 1. Introduce the topic and activity; suggestions are in the Teacher's Guide. 2. Divide students into groups of two or three and provide each group with the Student Puzzle Pieces and the Key to Wegener's Fossil Evidence. 3. Direct students to attempt to reassemble the land masses into one large land mass, Pangaea. 4. Follow the Teacher's Guide for complete lesson implementation instructions. 5. The guide provides several options for **assessment.**	spreading rates, and explain the significance of different spreading rates. The activity is *The Race is On . . . with Seafloor Spreading*, which can be accessed at http://joidesresolution.org/node/3152. Once they have completed the activity, the students will answer questions provided in the Teacher's Guide. Once they have completed the activity, as an **assessment**, the students will form their own opinions about whether the evidence they researched is compelling enough for them to accept the Theory of Plate Tectonics. Follow the steps below to complete the activity: 1. Introduce the topic and activity; suggestions are in the Teacher's Guide. 2. Divide students into groups of two or allow them to work alone. 3. Distribute the student handouts. 4. Direct students to study the handouts and complete the activity. 5. Follow the Teacher's Guide for complete lesson implementation instructions. 6. Engage the students in a discussion of their results.
Implementation	**Typical Students:** The time frame is one to two class periods, depending on how the teacher implements the activity. If this is students' first exposure to plate tectonics, you may want to show this YouTube video first: http://www.youtube.com/watch?v=_5q8hzF9VVE. Always preview any videos for appropriate content for your particular students.	

continued

Middle School: Earth's Systems, *continued*	
	Materials: Teacher Copy of Wegener's Key to Continental Positions, Teachers Guide, and Student Handouts (http://volcanoes.usgs.gov/about/edu/dynamicplanet/wegener/), This Dynamic Planet booklet (http://pubs.usgs.gov/gip/dynamic/); crayons, markers, or colored pencils; scissors; tape or glue; construction paper or poster board. **Advanced Students:** The time frame is one to two class periods. **Materials:** Student Handout pages, Teacher's Guide (http://joidesresolution.org/node/3152), rulers, pencils. **Extension:** For students who want more information about continental drift and plate tectonics, try the activity Plate Tectonics and Contributions from Scientific Ocean Drilling at http://joidesresolution.org/node/3150.
Foundation Box	**Science and Engineering Practices:** • Analyze and interpret data to provide evidence for phenomena. **Disciplinary Core Ideas:** • ESS2.B: Maps of ancient land and water patterns, based on investigations of rocks and fossils, make clear how Earth's plates have moved great distances, collided, and spread apart. **Crosscutting Concepts:** • Patterns: Patterns in rate of change and other numerical relationships can provide information about natural systems. • Nature of Science: Science findings are frequently revised and/or reinterpreted based on new evidence.
Connections to Common Core State Standards	**ELA/Literacy:** RST. 6–8.1: Cite specific textual evidence to support analysis of science and technical texts. RST.6–8.7: Integrate quantitative or technical information expressed with words in a text with a version of that information expressed visually (e.g., in a flowchart, diagram, model, graph, or table). RST. 6–8.9: Compare and contrast the information gained from experiments, simulations, videos, or multimedia sources with that gained from reading a text on the same topic. *continued*

Middle School: Earth's Systems, *continued*

Mathematics:

MP.2: Reason abstractly and quantitatively

6.NS.C.5: Understanding that positive and negative numbers are used together to describe quantities having opposite directions or values; use positive and negative numbers to represent quantities in real-world contexts, explaining the meaning of zero in each situation.

6.EE.B.6: Use variables to represent numbers and write expressions when solving a real-world or mathematical problem; understand that a variable can represent an unknown number or, depending on the purpose at hand, any number is a specified set.

7.EE.B.4: Use variables to represent quantities in a real-world or mathematical problem, and construct simple equations and inequalities to solve problems by reasoning about the quantities.

Note: Typical Activity adapted from *Wegner's Puzzling Evidence Exercise* by the United States Geological Survey, Department of the Interior, 2014. Adapted with permission. Advanced Activity is adapted from *The Race Is On . . . with Seafloor Spreading!* by The International Ocean Discovery Program, 2014. Adapted with permission.

Middle School
MS-ETS1
DCI: Engineering Design

Teacher Information: In this task, students are going to define the criteria and constraints of a design problem. They will not actually build the design at this point, but the design could be saved and used in other Engineering Design lessons that will focus on construction. Students will be thinking about how to design a tool, process, or system that can be used to assist students at their school who have physical disabilities. To meet this PE, students only need to define the criteria and constraints to such a degree that their solution would be potentially successful. The problem for the typical students has a narrow focus—that of solving a problem dealing with objects falling off a flat surface. The problem for the advanced students has a broad focus in that the students must come up with the specific problem they want to solve.

A great source for introductory videos on design is PBS Learning Media (http://www.pbslearningmedia.org/). Go to the "Science" section under "Subjects" and choose "Engineering," then search by clicking "Engineering Design." There are several videos on the design process at various grade levels. You might also be interested in the video about designing a wheelchair for a young lady to use to play rugby.

Note: Engineering design connects with several of the Performance Expectations. MS-ETS1.A, which is featured here, can also connect with MS-PS3-3. Thus an alternate lesson can be designed by the teacher using the same foundation boxes while meeting two Performance Expectations, one in physical science and one in engineering design.

Typical Learner(s)	Advanced Learner(s)
Essential Question: How can scientific knowledge and principles be used to design a tool, process, or system that will assist a student who is physically disabled to perform a task that the student cannot currently perform due to the disability?	
Directions: Explain the problem to the students: There are some students in our school and community who have specific physical disabilities that impede their ability to grasp and contain objects on a flat surface. Several of these students are in wheelchairs with trays attached. Others sit at tables using a restraint to hold them upright. If the student tries to grasp an object on *continued*	**Directions:** Explain the problem to the students: There are some students in our school and community who have specific physical disabilities that impede their ability to perform certain everyday tasks. Your job is to design a tool, process, or system that can be used to allow a student to perform a particular task that the student cannot currently per- *continued*

Performance Expectation:
MS-ETS1-1. Define the criteria and constraints of a design problem with sufficient precision to ensure a successful solution, taking into account relevant scientific principles and potential
continued

Middle School: Engineering Design, *continued*		
impacts on people and the natural environment that may limit possible solutions.	the table or tray and in the process, knocks the object to the floor, the student has to rely on someone else to retrieve the object. Your job is to design a tool, process, or system that can be used to allow the student to grasp the object on the tray and easily retrieve it if it gets knocked off. Typical objects that might be on the tray or table would be a cup, small ball or toy, stuffed animal, book, iPad, stylus, or pencil. You may think of other items. Choose an object as the focus of your project. Students may work alone or in small groups. Students will most likely need some structure for the project unless you have focused on this PE before. Here is a suggested list of prompts: 1. What are the criteria for the problem? 2. What other information do you need? Where might you find that information? 3. What scientific knowledge will you access to help you? What scientific principles? 4. What constraints might you have for solving the problem? 5. Are there any potential negative impacts on either people or the environment? If so, what are they? Could you devise a way to eliminate them? 6. Brainstorm ideas for solving the problem. 7. Choose a solution and describe it in words and diagrams. Explain any possible issues that would prevent you from following through with this solution.	form due to the student's specific physical disability. Students may work alone or in small groups. Students will most likely need some structure for the project unless you have focused on this PE before. Here is a suggested list of prompts: 1. What are the criteria for the problem? 2. What other information do you need? Where might you find that information? 3. What scientific knowledge will you access to help you? What scientific principles? 4. What constraints might you have for solving the problem? 5. Are there any potential negative impacts on either people or the environment? If so, what are they? Could you devise a way to eliminate them? 6. Brainstorm ideas for solving the problem. 7. Choose a solution and describe it in words and diagrams. Explain any possible issues that would prevent you from following through with this solution.

Middle School: Engineering Design, *continued*	
Implementation	Students should have their science notebooks available for note taking, sketching designs, and keeping track of their ideas, thinking, and **assessment** of their work. Have computers available for students to use in searching for information and for those who might want to use a drawing program to sketch their designs. **Assessment** would include examining the science notebooks for evidence that a procedure was followed and that students elaborated on the prompts. In addition, students should have a detailed diagram with accompanying written explanation for their solution. As an extension, students could work toward constructing their tool, process, or system.
Foundation Box	**Science and Engineering Practices:** • Asking Questions and Defining Problems **Disciplinary Core Ideas:** • ETS1.A: Defining and Delimiting Engineering Problems **Crosscutting Concepts:** • Influence of Science, Engineering and Technology on Society and the Natural World
Connections to Common Core State Standards	**ELA/Literacy:** **RST. 6–8.1:** Cite specific textual evidence to support analysis of science and technical texts. **WHST.6–8.8:** Gather relevant information from multiple print and digital sources; assess the credibility of each source; quote or paraphrase the data and conclusions of mothers while avoiding plagiarism and providing bibliographic information for sources. **Mathematics** **MP.2:** Reason abstractly and quantitatively **7. EE.B.3:** Solve multi-step, real-world and mathematical problems posed with positive and negative rational numbers in any form, using tools strategically. Apply properties of operations to calculate with numbers in any form; convert between forms as appropriate; assess the reasonableness of numbers using mental computation and estimation strategies.

High School
HS-PS1-1
DCI: Structures and Properties of Matter: Matter and its Interactions

Teacher Information: This lesson focuses on the repeating patterns of the periodic table, which reflect the patterns of outer level electrons. Students should be able to determine an element's electron configuration and valence electrons. Students should be able to explain various chemical and physical properties of elements including ionization energy, reactivity of metals, types of bonds formed, and so on. The teacher should determine through preassessment, observation, or formative assessment which students are ready for the advanced learning experience. Advanced learners will use their knowledge of periodic patterns to predict properties of an unknown element.

Typical Learner(s)	Advanced Learner(s)
Essential Question: How are the relative properties of elements determined by their location on the periodic table?	

Typical Learner(s)	Advanced Learner(s)
Directions: Engage students by allowing them to observe the reactions of Group 1 metals with water. Due to safety concerns, this is best done by viewing a video. These can be easily found on YouTube.	**Directions:** Engage students by allowing them to observe the reactions of Group 1 metals with water. Due to safety concerns, this is best done by viewing a video. These can be easily found on YouTube.
After watching the video, divide students into groups and instruct each group to research the properties of one Group 1 metal. Using the jigsaw method, have students share their findings.	After watching the video, divide students into groups and instruct each group to research the properties of one Group 1 metal. Using the jigsaw method, have students share their findings.
As a whole group, encourage students to recall that Group 1 metals have one valence electron. Have students predict properties of other groups based on their valence electrons. Inform students that many properties of elements can be predicted based on their location on the periodic table. Explain how Mendeleev was able to predict the properties of elements that were not discovered based on the gaps that were left in the table. *continued*	As a whole group, encourage students to recall that Group 1 metals have one valence electron. Have students predict properties of other groups based on their valence electrons. Inform students that many properties of elements can be predicted based on their location on the periodic table. Explain how Mendeleev was able to predict the properties of elements that were not discovered based on the gaps that were left in the table. *continued*

Performance Expectation:
HS-PS1-1. **Use the periodic table as a model to predict the relative properties of elements based on the patterns of electrons in the outermost energy level of atoms.**
Clarification Statement: Examples of properties that could be predicted from patterns could include reactivity of metals, types of bonds formed, numbers of bonds formed, and reactions with oxygen.
Assessment Boundary: Assessment is limited to main group elements. Assessment does not include quantitative understanding of ionization energy beyond relative trends.

High School: Structures and Properties of Matter, *continued*

	Students will create a 3D Periodic Table noting the periodic trends. This Flinn Scientific activity can be found at this link: http://cationclub.files.wordpress.com/2013/10/cf10480.pdf. In a culminating activity, students will be given 15 mystery elements and be asked to arrange them in a table according to their chemical and physical properties. Provide students with a list of the properties for each unknown element such as: melting and boiling point, ionization energy, atomic radius, electron affinity, and electronegativity.	Students will create a 3D Periodic Table noting the periodic trends. This Flinn Scientific activity can be found at this link: http://cationclub.files.wordpress.com/2013/10/cf10480.pdf. In a culminating activity, students will be given 15 mystery elements and be asked to arrange them in a table according to their chemical and physical properties. Provide students with a list of the properties for each unknown element such as: melting and boiling point, ionization energy, atomic radius, electron affinity, and electronegativity. In addition to doing the activity, learners will predict the properties (melting and boiling point, ionization energy, atomic radius, electron affinity, and electronegativity) of the elements surrounding the 15 unknown elements. To do this, students will need to plot the periodic properties of the known elements and extrapolate the unknown quantities.
Implementation	**Materials:** index cards, seven 96–well reaction plates (8 x 12 layout), periodic table, straws, scissors, metric rulers, clay. This lesson should take at least two (45-minute) class periods. **Assessment note:** Students should recognize and explain the periodic trends as displayed in the 3D periodic table. In addition, students should be able to predict properties of elements based on their location on the periodic table.	

High School: Structures and Properties of Matter, *continued*

Foundation Box	**Science and Engineering Practices:** • Developing and Using Models **Disciplinary Core Ideas:** • PS1.A: Structure and Properties of Matter **Crosscutting Concepts:** • **Patterns:** Different patterns may be observed at each of the scales at which a system is studied and can provide evidence for causality in explanations of phenomena. • **Nature of Science:** Science assumes the universe is a vast single system in which basic laws are consistent.
Connections to Common Core State Standards	**ELA/Literacy:** **RST.9–10.7:** Translate quantitative or technical information expressed in words in a text into visual form (e.g., a table or chart) and translate information expressed visually or mathematically (e.g., in an equation) into words.

Note: Typical and advanced activitiy from *Plotting Trends: A Periodic Table Activity* by Flinn Scientific, Inc., 2006, Batavia, IL: Flinn Scientific, Inc. Copyright 2006 by Flinn Scientific Inc. Used with permission.

High School
HS-LS1
DCI: From Molecules to Organisms: Structure and Processes

Teacher Information: Typical learners should have a good understanding of the relationship between cells, tissues, organs, and organ systems and how they contribute to the functions of life. Typical learners should also be able to design an experiment with controlled variables to answer a scientific question. Advanced learners should already have an understanding of homeostasis at the organism level, such as how the human body responds to cold temperatures.

	Typical Learner(s)	Advanced Learner(s)
	Essential Question: How do organisms maintain internal stability in an ever-changing environment?	
Performance Expectation: **HS-LS1-3. Plan and conduct an investigation to provide evidence that feedback mechanisms maintain homeostasis.** *Clarification Statement:* Examples of investigations could include heart rate response to exercise, stomate response to moisture and temperature, and root development in response to water levels. *Assessment Boundary:* Assessment does not include the cellular processes involved in the feedback mechanism.	**Directions:** *Whole-group engagement:* Ask students how they feel. Have students describe how they know they are ill. Instruct students to brainstorm ways in which the body responds to illness, changes in the environment, and so on. *Small-group exploration:* Instruct students to design an experiment to test an organism's response to a changing environment. (This could be as simple as investigating the effects of exercise on heart rate or investigating a plant mechanism such as stomate response to environmental factors.) Conclude the experiment with students, describing how these responses help maintain stability or homeostasis in the organism. Students should distinguish between the types of homeostatic system observed during their experiment (e.g., thermoregulation, osmoregulation, waste management). *continued*	**Directions:** *Whole-group engagement:* Ask students how they feel. Have students describe how they know they are ill. Instruct students to brainstorm ways in which the body responds to illness, changes in the environment, and so on. *Small-group exploration:* Students will diverge from the typical learner during the small-group exploration. Direct students to design an experiment that provides more detail with regard to feedback mechanisms and how cells, tissues, and organs of the affected organ system respond. For example, advanced students should explain how exercise stimulates an increase in heart rate, respiratory rate, temperature, etc. and what occurs at the cellular level that makes these changes possible. *continued*

High School: From Molecules to Organisms, *continued*

	Small-group elaboration: Provide students with case studies to analyze for homeostatic feedback mechanisms. Example case studies and questions can be found at the following websites or by searching for homeostasis case studies: • http://dustinsmerekstudentteaching.webs.com/Homeostasis_SMEREK.pdf • http://teacherweb.com/NJ/QueenofPeaceHighSchool/McGill/ACaseStudy–Hyperther.doc	*Small-group elaboration:* Advanced students should be given a more complex case study and again, describe the responses with regard to cells, tissues, and organs. An example can be found at The National Center for Case Study Teaching in Science: http://science cases.lib.buffalo.edu/cs/collection/detail.asp?case_id=366&id=366. Specifically, *The 2000–Meter Row* by Nathan Strong allows students to recognize how organ systems are related in the process of homeostasis.
Implementation	This lesson would take at least 2, possibly 3 days to complete. Students may need a day to design their experiment and a day to collect data and draw conclusions. **Assessment note:** The case studies could be used as a pre– and postassessment. Students who identify and recognize homeostatic mechanisms in the case studies could further their study in the advanced group. Additionally, students should be assessed on their conclusions from the investigation. Students should distinguish between the types of homeostatic system observed during their experiment (e.g., thermoregulation, osmoregulation, waste management).
Foundation Box	**Science and Engineering Practices:** • Planning and Carrying Out Investigations **Disciplinary Core Ideas:** • LS1.A: Structure and Function

continued

High School: From Molecules to Organisms, *continued*	
	Crosscutting Concepts: • **Stability and Change:** Feedback (negative or positive) can stabilize or destabilize a system. • **Nature of Science:** Scientific inquiry is characterized by a common set of values that include: logical thinking, precision, open-mindedness, objectivity, skepticism, reliability of results, and honest and ethical reporting of findings.

Connections to Common Core State Standards	**ELA/Literacy:**
	   WHST.9–12.7: Conduct short as well as more sustained research projects to answer a question (including a self-generated question) or solve a problem; narrow or broaden the inquiry when appropriate; synthesize multiple sources on the subject, demonstrating understanding of the subject under investigation.
	   WHST.11–12.8: Gather relevant information from multiple authoritative print and digital sources, using advanced searches effectively; assess the strengths and limitations of each source in terms of the specific task, purpose, and audience; integrate information into the text selectively to maintain the flow of ideas, avoiding plagiarism and overreliance on any one source and following a standard-format for citation.

High School HS-ESS3-5 DCI: Weather and Climate: Global Climate Change	**Teacher Information:** To capture students' attention and engage them in active learning, the lesson will begin with an activity modeling the effects of carbon dioxide on temperature. Typical learners should be able to distinguish between weather and climate. They should recognize that weather and climate are influenced by geological interactions of sunlight, the ocean, the atmosphere, ice, landforms, and living things. Students should be able to identify how human activities such as burning of fossil fuels contribute to excess carbon dioxide and other pollutants in the atmosphere. Using climate forecasting data, advanced students will predict the effects of climate changes on local biomes and suggest possible methods to prevent the negative effects. The teacher should determine through preassessment, observation, or formative assessment which students are ready for the advanced learning experience.	
	Typical Learner(s)	**Advanced Learner(s)**
Performance Expectation: **HS-ESS3-5 Analyze geoscience data and the results from global climate models to make an evidence-based forecast of the current rate of global or regional climate change and associated future impacts to Earth's systems.** *Clarification Statement:* Examples of evidence, for both data and climate model outputs, are for climate changes (such as precipitation and temperature) and their associated impacts (such as on sea level, glacial ice volumes, and atmosphere and ocean composition). *continued*	**Essential Question:** How is Earth's climate changing and what environmental impacts can be expected?	
	Directions: Students will participate in an activity that models the effects of carbon dioxide gas on temperature. One such experiment can be found on the PBS website: http://www.pbs.org/wgbh/nova/educa tion/viewing/0302_03_nsn.html. Next, students are divided into groups and will research climate changes (precipitation and temperature) and their associated impacts. Students will explain how human activities have contributed to these changes. Teachers should try to extend student thinking using the results of the previous lab investigation. Using student climate data from a site such as University of New Hampshire Student Climate Data (http://studentclimatedata.unh.edu/index.shtml), students will observe the past and future trends in climate *continued*	**Directions:** Students will participate in an activity that models the effects of carbon dioxide gas on temperature. One such experiment can be found on the PBS website: http://www.pbs.org/wgbh/nova/educa tion/viewing/0302_03_nsn.html. Next, students are divided into groups and will research climate changes (precipitation and temperature) and their associated impacts. Students will explain how human activities have contributed to these changes. Teachers should try to extend student thinking using the results of the previous lab investigation. After completing the two initial activities, advanced students will predict future climate changes on a local biome using the climate forecasting data. Students should predict the effects on the ecosystem of the biome *continued*

High School: Weather and Climate, *continued*

Assessment Boundary: Assessment is limited to one example of a climate change and its associated impacts.	change in a geographic location of their choice and predict at least two environmental effects from these changes. Students could use various presentation methods to present findings to the class.	of their choice, including how the changes might impact human activities. Students will present information to the class and propose a method to prevent further negative impacts. The following site can be used as a guide for this lesson: http://studentclimatedata.unh.edu/climate/teacher_resources/ResearchPlanningGuide_TeacherVersion.pdf.
Implementation	**Materials:** Internet access, two or more 2-liter clear soda bottles with the label removed, identical thermometers for each soda bottle, opaque tape, source of carbon dioxide (CO_2), spray paint or spray glitter, modeling clay. Be sure to follow science safety rules when conducting the experiment. Prior to the lesson, make sure you are able to access the climate forecasting data from a website such as http://studentclimatedata.unh.edu/index.shtml. Students should be allowed to have choices in their presentation mode (e.g., PowerPoint, Prezi, Voki). **Assessment note:** Students should be assessed on their prediction of environmental effects using climate data. The assessment would include how well students are able to support their conclusions with evidence from the climate data. This lesson should take at least three (45-minute) class periods.	
Foundation Box	**Science and Engineering Practices:** • Analyzing and Interpreting Data **Disciplinary Core Ideas:** • ESS3.D: Global Climate Change	

continued

High School: Weather and Climate, *continued*

	Crosscutting Concepts: • **Stability and Change:** Change and rates of change can be quantified and modeled over very short or very long periods of time. Some system changes are irreversible. • **Nature of Science:** Science investigations use diverse methods and do not always use the same set of procedures to obtain data. New technologies advance scientific knowledge. Science knowledge is based on empirical evidence. Science arguments are strengthened by multiple lines of evidence supporting a single explanation.
Connections to Common Core State Standards	**ELA/Literacy:** **RST.11–12.1:** Cite specific textual evidence to support analysis of science and technical texts, attending to important distinctions the author makes and to any gaps or inconsistencies in the account. **RST.11–12.2:** Determine the central ideas or conclusions of a text; summarize complex concepts, processes, or information presented in a text by paraphrasing them in simpler but still accurate terms. **RST.11–12.7:** Integrate and evaluate multiple sources of information presented in diverse formats and media (e.g., quantitative data, video, multimedia) in order to address a question or solve a problem. **Mathematics:** **MP.2:** Reason abstractly and quantitatively. **HSN.Q.A.1:** Use units as a way to understand problems and to guide the solution of multistep problems; choose and interpret units consistently in formulas; choose and interpret the scale and the origin in graphs and data displays. **HSN.Q.A.2:** Define appropriate quantities for the purpose of descriptive modeling. **HSN.Q.A.3:** Choose a level of accuracy appropriate to limitations on measurement when reporting quantities.

High School
HS-ETS1-2
DCI: Engineering Design

Performance Expectation:
HS-ETS1-2. Design a solution to a complex real-world problem by breaking it down into smaller, more manageable problems that can be solved through engineering.

HS-ETS1-3. Evaluate a solution to a complex real-world problem based on prioritized criteria and trade-offs that account for a range of constraints including cost, safety, reliability, and aesthetics as well as possible

continued

Teacher Information: In this task, students are going to design and evaluate a solution to a complex real-world problem. This could be an activity used with the study of water or water purification. Students should recognize the properties of water and the importance of water to human existence. Students should evaluate resources that can be used to design a solution to the problem. Advanced students should be capable of solving more complex problems with a greater number of variables.

Note: Engineering design connects with several of the Performance Expectations. HS-ETS1-2, which is featured here, can also connect with HS-ESS3.C. Thus an alternate lesson can be designed by the teacher using the same foundation boxes while meeting two Performance Expectations, one in earth and space science and one in engineering design.

Essential Question: How do living things adapt to their environment for survival?

Typical Learner(s)	Advanced Learner(s)
Directions: Present students the following scenario: Your plane crashed on the *Lost* island somewhere in the tropics. There are 20 survivors with no way to contact outside help. A waterfall was found about 2 miles from the beach. You could move your camp to the area near the waterfall but you need to be easily seen by potential rescuers. In addition, some of the people are frightened by a loud, moving band of smoke that lurks in the jungle. Your task is to design a structure to transport large amounts of water from the waterfall to the beach. The following materials have been gathered for your use from the jungle and wrecked airplane: • Water bottles • Plastic drop cloth • Buckets *continued*	**Directions:** Present students the following scenario: Your plane crashed on the *Lost* island somewhere in the tropics. There are 20 survivors with no way to contact outside help. A waterfall was found about 2 miles from the beach. You could move your camp to the area near the waterfall but you need to be easily seen by potential rescuers. In addition, some of the people are frightened by a loud, moving band of smoke that lurks in the jungle. Your task is to design a structure to transport large amounts of water from the waterfall to the beach. The following materials have been gathered for your use from the jungle and wrecked airplane: • Water bottles • Plastic drop cloth • Buckets *continued*

High School: Engineering Design, *continued*

social, cultural, and environmental impacts. **HS-ESS3-1. Construct an explanation based on evidence for how the availability of natural resources, occurrence of natural hazards, and changes in climate have influenced human activity.** **ESS2.C: The Roles of Water in Earth's Surface Processes.** **The abundance of liquid water on Earth's surface and its unique combination of physical and chemical properties are central to the planet's dynamics. These properties include water's exceptional capacity to absorb, store, and release large amounts of energy, transmit sunlight, expand upon freezing, dissolve and transport materials, and lower the viscosities and melting points of rocks.** *continued*	• Bamboo trees (clear vinyl tubing) • Lumber from trees in the jungle • Twine (*Note:* Activity can be accessed at: http://www.teachengineering.org/view_activity.php?url=collection/wpi_activities/wpi_construct_an_aqueduct/construct_an_aqueduct.xml#mats) To evaluate the best possible solutions, student groups could compete against each other to see who was able to gather the most water in the shortest amount of time.	• Bamboo trees (clear vinyl tubing) • Lumber from trees in the jungle • Twine (*Note:* Activity can be accessed at: http://www.teachengineering.org/view_activity.php?url=collection/wpi_activities/wpi_construct_an_aqueduct/construct_an_aqueduct.xml#mats) To evaluate the best possible solutions, student groups could compete against each other to see who was able to gather the most water in the shortest amount of time. Advanced students can be presented more obstacles, such as: • Add blocks perpendicular to the water flow to represent small hills. Additionally, advanced students could be asked to design a water purification station using the materials below. You could suggest the water has been contaminated and the students would need to design a solar water still to make the water drinkable. • Nails/screws (from the airplane) *continued*

High School: Engineering Design, *continued*

Clarification Statement: Examples of key natural resources include access to fresh water (such as rivers, lakes, and groundwater), regions of fertile soils such as river deltas, and high concentrations of minerals and fossil fuels. Examples of natural hazards can be from interior processes (such as volcanic eruptions and earthquakes), surface processes (such as tsunamis, mass wasting, and soil erosion), and severe weather (such as hurricanes, floods, and droughts). Examples of the results of changes in climate that can affect populations or drive mass migrations include changes to sea level, regional patterns of temperature and precipitation, and the types of crops and livestock that can be raised.	• Glass jar • Coffee Can • String or twine • Wood • Aluminum foil (sheet metal from the plane) • Glass • Clear plastic • Resin from rubber tree (caulk) Or . . . if particulates are present in the water, students can build a model water filtration system using the following materials: • sand • plastic jugs (airplane) • plastic straws (airplane) • coal (charcoal briquettes) • Cotton (airplane) • Masking tape (airplane) • Resin from rubber tree (hot glue gun and glue) • Fabric (airplane) (*Note:* Activity can be accessed at: http://cied.uark.edu/2009_UA_PROBLEM_SOLVING.pdf)
Implementation	**Materials:** Materials: water bottles, plastic drop cloth, buckets, bamboo trees (clear vinyl tubing), lumber from trees in the jungle (chairs and stools from classroom), twine, nails/screws (from the airplane), glass jar, coffee can, wood, aluminum foil (sheet metal from the plane), clear plastic, resin from rubber tree (caulk), sand, plastic

continued

High School: Engineering Design, *continued*	
	jugs (airplane), plastic straws (airplane), coal (charcoal briquettes), cotton (airplane), paper and cardboard stock (airplane), masking tape (airplane), resin from rubber tree (hot glue gun and glue), fabric (airplane). Set up the watercourse using a table with a child's pool on top and a bucket 5 feet away.
	Assessment note: Assessment can be based on how well students met the engineering design challenge following the specified constraints. Additional assessments could include students' science notebooks. Teachers can have students journal each step of the engineering design process, including sketching diagrams, noting successes and setbacks, and recording the procedure that was followed.
Foundation Box	**Science and Engineering Practices:** • Constructing Explanations and Designing Solutions **Disciplinary Core Ideas:** • HS-ESS3.C: Human Impacts on Earth Systems • HS-ETS1.B: Developing Possible Solutions • HS-ESS3-1: Construct an explanation based on evidence for how the availability of natural resources, occurrence of natural hazards, and changes in climate have influenced human activity. • HS-ESS2.C: The Roles of Water in Earth's Surface Processes **Crosscutting Concepts:** • **Systems and System Models:** When investigating or describing a system, the boundaries and initial conditions of the system need to be defined and their inputs and outputs analyzed and described using models. • **Influence of Science, Engineering, and Technology on Society and the Natural World:** New technologies can have deep impacts on society and the environment, including some that were not anticipated. Analysis of costs and benefits is a critical aspect of decisions about technology.

High School: Engineering Design, *continued*	
Connections to Common Core State Standards	**ELA/Literacy:** **RST.11–12.7:** Integrate and evaluate multiple sources of information presented in diverse formats and media (e.g., quantitative data, video, multimedia) in order to address a question or solve a problem. **RST.11–12.8:** Evaluate the hypotheses, data, analysis, and conclusions in a science or technical text, verifying the data when possible and corroborating or challenging conclusions with other sources of information. **RST.11–12.9:** Synthesize information from a range of sources (e.g., texts, experiments, simulations) into a coherent understanding of a process, phenomenon, or concept, resolving conflicting information when possible. **Mathematics:** **MP.2:** Reason abstractly and quantitatively. **MP.4:** Model with mathematics.

Note: Lost activity adapted from *Do as the Romans: Construct an Aqueduct!* by TeachEngineering, 2013, Denver: University of Colorado, TeachEngineering. Copyright 2013 by Regents of the University of Colorado. Adapted with permission.

Water purification activity adapted from *A Collection of Engineering Design Problems* (pp. 19–20), by T. Blunier, M. K. Daugherty, and L. Morford, 2009, Fayetteville: University of Arkansas. Adapted with permission.

Chapter 4

Problem- and Project-Based Learning

An instructional option that helps provide creativity and challenge is problem-based learning (PBL). Finkle and Torp (1995) described problem-based learning as a " . . . curriculum development and instructional system that simultaneously develops both problem solving strategies and disciplinary knowledge bases and skills by placing students in the active role of problem solvers confronted with ill-structured problems that mirror real-world problems" (p. 1). PBL has its roots in the medical field, used to provide medical students with simulations focused on situations that typically arise in one's practice at the hospital or in surgery (Gallagher, Sher, Stepien, & Workman, 1995). These researchers adapted it for use in science classes in elementary and high school settings. In addition to integrating PBL and science, the components of all problem-based episodes include:

> . . . initiating learning with an ill–structured problem, using the problem to structure the learning agenda, and teacher as metacognitive coach, with important goals of a reformed science curriculum such as learning based on con-

DOI: 10.4324/9781003238522-4

cepts of significance, student-designed experiments, and development of scientific reasoning
skills. (p. 137)

As an extension of constructivist thinking (Dewey, 1938;
Piaget, 1937; Vygotsky, 1978), PBL promotes an environment
in which ". . . learners are actively engaged in working at tasks
and activities which are authentic to the environment in which
they can be used" (Savery & Duffy, 1995, p. 13). In the field of
gifted education, intervention studies involving PBL units in
K–12 schools have demonstrated gains in student learning. For
example, VanTassel-Baska and Bass (1998) and VanTassel-Baska,
Avery, Hughes, and Little (2000) both examined science units
utilizing PBL. Both found significant gains in student learning,
particularly of the scientific process.

In the PBL environment, learning is student centered, and
students are typically arranged in groups. Students are presented
real-world problems that may or may not be presented as a fictitious scenario. To bring authenticity to a PBL, the problem may
be tied to an authentic community-based issue, and when acted
upon, can serve as a service-learning project. Because of similar
processes of investigation, PBL is complementary to the scientific
method of inquiry. Figure 4.1 highlights these similarities and
subtle differences.

In addition to alignment to the scientific method of inquiry,
advantages of problem-based learning include the development
of high levels of critical and scientific thinking, time management, data collection, report preparation, and evaluation skills
(Akinoglu & Tandogan, 2007). According to Cotabish et al.
(2014), PBL may increase habits of mind, as students act as real
scientists to find possible solutions to real-world problems. The
authors elaborate by stating, "Many science units can be reorga-
nized as PBL by creating a problem statement, providing students
with some background knowledge, then providing plenty of time
for them to conduct research, especially scientific experimenta-
tion" (p. 16). As an alternative to cookbook science often carried

Problem-Based Learning	Scientific Methods of Inquiry
Process of investigation	Process of investigation
Solves a genuine problem	Solves a genuine problem
Students analyze a problem	Observation/Ask a question
Conduct research (the three What's)	Conduct research
Define problem	Construct a hypothesis
Problem confirmed	Test the hypothesis
Propose a solution	Draw a conclusion
Problem solved/report results	Report results

Figure 4.1. Similarities between PBL and scientific method of inquiry.

out in schools, PBL reflects the demands of 21st-century skills made possible by using modes of open inquiry such as Herron's (1971) Model of Inquiry (see Figure 4.2).

Herron's model classifies inquiries according to a scale ranging from 0 (Confirmation/Verification) to 3 (Open Inquiry). The level of inquiry is dependent upon teacher support and whether there is an existing solution to the problem or question. It is important to note that most students will need to be provided some level of teacher support before moving entirely to open inquiry. Herron's (1971) Model of Inquiry aligns well to scientific methodology and lends itself to gifted education pedagogy.

Typically, there are six steps in conducting problem-based learning; however, steps two through five are not necessarily sequential and may be conducted simultaneously, as new information may redefine the original problem. In addition, we propose a seventh (optional) step, resolution/action, because an additional step may be necessary to carry out the solution. These seven steps are:

1. Introduce an ill-structured problem.
2. Identify the three "What's" also known as the "Need to Know's." Students list what is known, what they need to know, and what they need to do.
3. Gather information.
4. Brainstorm possible solutions.

Level of Inquiry	Question	Procedure	Solution
Confirmation	Provided	Provided	Provided
Structured	Provided	Provided	Student Generated
Guided	Provided	Student Generated	Student Generated
Open	Student Generated	Student Generated	Student Generated

Figure 4.2. Herron's Model of Inquiry.

5. Determine the solution that is the best fit.
6. Present the solution.
7. Resolution/Action (optional)

To demonstrate how PBL can be carried out in a gifted classroom, we will highlight a PBL science experience in action.

Step 1: The Ill-Structured Problem

In 2011, Ms. Kati Searcy's fifth-grade gifted and talented class at Mountain Park Elementary (Fulton County Schools in Roswell, GA) decided to adopt Rocky Creek, a creek located across from the school. Students noticed that there was extreme erosion on the banks of Rocky Creek. As a learning experience tied to a community service project, Ms. Searcy assigned her class to approach the problem through PBL. Using the seven-step process, students were asked to carry out the investigation of Rocky Creek and propose a realistic solution to be acted upon by the group.

Step 2: Identify the Three "What's"

The Three "What's"

What We Know	What We Need to Know	What We Need To Do
There is considerable erosion at the local creek. Many roots of huge trees are exposed. Trees will possibly fall.	What exactly will happen if the bank continues to erode? What will happen if the large trees come down? Can the erosion be stopped or slowed down? If so, how? Why isn't the city doing anything to help stabilize the bank?	Research creek erosion. Contact the city's environmental department. Talk to a specialist on native trees and plants.

Step 3: Gather Information

To gather information, students contacted the city's environmental division and an environmental specialist. Furthermore, students researched erosion including the causes leading to erosion and possible solutions to remedy the problem.

Step 4: Possible Solutions

After gathering information, students generated a list of possible solutions to remedy the problem.

Possible Solutions

Post "Danger" signs at the creek.
Plant trees and shrubs to help stop the erosion.
Do nothing—you can't stop erosion. It's a natural process.
Ask the city to provide canvas shelters from one tree to another above the eroding banks.

Step 5: Determine Best Fit Solution

As a tool for problem solving, students created a table to assess their options and analyzed the data to determine the solution that was the best fit.

Finding the Best Solution

	Will help the problem?	We can do this?	Easy to do?	Number of "yes" responses
Post "Danger" signs	No	Yes	Yes	2
Plant trees and shrubs	Yes	Yes	Yes	3
Do nothing	No	Yes	Yes	2
City provides canvas shelters	Yes	No	Yes	1

Step 6: Present the Solution

After determining that planting trees and shrubs was the best possible solution to resolve the problem (based on the results in Step 5), students found that they needed additional information and generated a list of questions.

Additional Questions

What trees and shrubs are native to the area?

How much will it cost to plant native trees and shrubs?

How will we raise the money to buy the native trees and shrubs?

Step 7: Resolution/Action (Optional)

In this particular case, Ms. Searcy's class decided to act on the solution. In order to accomplish the task of raising money to buy trees and shrubs to prevent further bank erosion, Ms. Searcy's class formed a company called "Adopt-a Wreath." Students made and sold products, created commercials to market their products, and used the funds to purchase the trees and shrubs. For their efforts, these gifted students were presented with the *Georgia Adopt-A-Stream Action Award.*

As illustrated in this example, problem-based learning allows a teacher many opportunities for meeting the needs of diverse learners, including gifted learners. Within the context of collaboration, a 21st-century skill, students often work in teams and independently as they engage in inquiry. This allows teachers to provide differentiated support to teams and individuals. In addition, the example illustrates how to address the NGSS Discipline Core Idea of Earth's systems (4-ESS2). Furthermore, Crosscutting Concepts such as cause and effect relationships as well as the Science and Engineering Practices of planning and carrying out investigations to test solutions are all evident in Ms. Searcy's problem-based learning activity. Their use of PBL to solve a real-world ecology problem along with their innovative approach to resolve the issue with action was a successful demonstration of PBL in use.

Project-Based Learning

> If we are serious about reaching 21st century education goals, project-based learning must be at the center of 21st century instruction. The project contains and frames the curriculum, which differs from the "short" project or activity added onto traditional instruction. Project-based learning is the *Main Course*, not the *Dessert.*"
> —Buck Institute for Education (n.d., para. 1)

Similar to problem-based learning, project-based learning is an instructional strategy in which students work cooperatively over time to create a product, presentation, or performance. The fundamental difference between problem- and project-based learning is in the application. Both provide opportunities for differentiation, are open-ended in nature, follow a list of steps, and utilize inquiry for the basis of student-centered instruction; however, there are some differences to note. Project-based learning is always grounded in an authentic, real-world problem, whereas problem-based learning may use a fictitious scenario to present a problem. Project-based learning experiences always culminate in the presentation of a product compared to problem-based learning experiences, which may result in a solution rather than a product. In project-based learning, students may be involved in the development of rubrics, through which assessment of outcomes, including student self-assessment, may be conducted. Like problem-based learning, project-based learning is grounded in 21st-century skills, engages students in metacognitive thinking, and may likely improve scientific habits of mind and higher order thinking. Figure 4.3 presents the similarities and differences between the two PBLs.

Project-based learning requires students to respond to a complex question, problem, or challenge, and is guided by a set of principles that guide teacher and student experiences. Like problem-based learning, teachers act as facilitators, guiding student instruction and experiences. Students are expected to contribute to the learning process as well as collaborate with classmates in their quest to produce artifacts that are generated through the use of technology (Colley, 2008). Unlike add-on projects, project-based learning experiences are grounded in context utilizing an essential/driving question to frame the learning experience. Within the project, the teacher scaffolds the learning for students (e.g., labs, lectures, technology applications, instructional activities). Like problem-based learning, students generate a list of three "What's" or "Need to Know's" in order to proceed with the project. The application of these elements differentiates project-based learning from other learning experiences. According to the Buck Institute (n.d.), essential elements of project-based learning include:

- **Significant content:** At its core, the project is focused on teaching students important knowledge and skills, derived from standards and key concepts at the heart of academic subjects.

Similarities	
Both: • Provide opportunities for differentiation • Are open-ended in nature • Address 21st-century learning competencies • Are task driven • Employ entry events • Typically are conducted in groups • Are student centered • Used as a formative assessment • Includes the three "What's" or "Need to Know's" • Involve research of subject matter • Spur in-depth inquiry • Follow steps • Prompt critical and creative thinking	

Differences	
Problem-based learning	**Project-based learning**
Typically shorter in duration	Often longer in duration
Choice is tied to possible solutions	Frequently employs student choice throughout
Often single subject	Often interdisciplinary/integrative
Products are often in the form of solutions	Emphasis on final product
Multiple paths for solving ill-structured problem	Centered around driving questions
Newfound information may redirect or pose additional questions	Final products are often presented to public audiences
Often uses case studies or fictitious scenarios to set up the problem	Typically involves real-world problem
May require an additional action step to carry out and resolve the issue(s)	Employs revision and reflection
May or may not utilize technology	Utilizes technology

Figure 4.3. Problem-based versus project-based learning.

- **21st-century competencies:** Students build competencies valuable for today's world, such as critical thinking/problem solving, collaboration, communication, and creativity/innovation, which are taught and assessed.

- **In-depth inquiry:** Students are engaged in a rigorous, extended process of asking questions, using resources, and developing answers.
- **Driving question:** Project work is focused by an open-ended question that students understand and find intriguing, which captures their task or frames their exploration.
- **Need to know:** Students see the need to gain knowledge, understand concepts, and apply skills in order to answer the driving question and create project products, beginning with an entry event that generates interest and curiosity.
- **Voice and choice:** Students are allowed to make some choices about the products to be created, how they work, and how they use their time, guided by the teacher and depending on age level and project-based learning experience.
- **Revision and reflection:** The project includes processes for students to use feedback to consider additions and changes that lead to high-quality products and to think about what and how they are learning.
- **Authentic audience:** Students present their work to other people, beyond their classmates and teacher.

Science, technology, engineering, and mathematics (STEM) education can be easily integrated with project-based learning. In the NGSS, the performance expectations connect content, big ideas in science, and engineering practices. As the centerpiece of instruction, science content comes to life through a project-based learning framework that includes science content, practice, essential questions and understandings, and applies them to problems that will meaningfully engage students.

With regard to advanced learners, project-based learning experiences lend themselves to student ownership through student choice and personal interest, and there are many opportunities to provide appropriate levels of challenge. Project-based learning experiences can be easily differentiated, offer opportunities for acceleration of content, and promote problem solving, collaboration, and critical thinking. Advanced learners are not limited by "ceilings," and the use of technology applications provide opportunities for gifted students to excel without imposed limitations. Additionally, gifted students are often involved in the assessment process, creating their own rubrics and monitoring plans tied to individualized project goals. Figure 4.4 presents the relationships between project-based learning, the NGSS, and gifted education pedagogical practices.

To demonstrate how project-based learning can be carried out, let us consider a common bridge design challenge. Typically, students are asked to design an effective bridge with toothpicks or digitally by using software. To make it a project-based learning experience, teachers would want to map out learning goals, consider how students will demonstrate what they have learned, craft a driving question, consider assessment strategies, and plan for a culminating event. Good entry events (scenarios) set the stage for the learning experience. Entry events should be novel and memorable. For example, an entry event for a bridge design challenge could be:

> City officials want to extend a bike trail across the river. A two-lane bridge exists, but a bike path is not part of the current design. Students are to design a bridge or redesign the existing bridge in order to accommodate a bike trail, and present their design to city officials and/or architectural engineers.

To increase student interest in the entry event, teachers may want to utilize video clips from websites such as YouTube, provide a simulation activity from an interactive science website, conduct a live or virtual field trip, invite a guest speaker or expert to introduce the entry event, include a primary or secondary

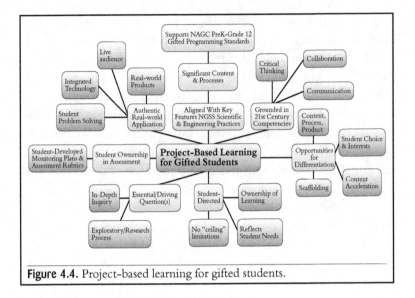

Figure 4.4. Project-based learning for gifted students.

source document, or showcase music or artwork to support the entry event.

Combined with the driving (essential) question, the entry event spurs the "Need to Know's." A **driving question** for the current example could be:

> What makes a bridge strong? How do you design
> a sturdy bridge?

Throughout the project, project-based learning components will enable students to answer their questions through research, community outreach, and internal and external resources (e.g., consulting with architects, bridge experts, engineers, or city officials). Lastly, students' final product or performance will demonstrate what they have learned through the process (e.g., students could make recommendations for retrofitting a local bridge to accommodate the bike trail and present this information to city officials and/or architectural engineers). Regardless of the project-based learning experience, the teacher can differentiate the content, process, or product, or add complexity to the activity in order to ensure challenge for gifted students. In the current example, the teacher could add complexity and build upon engineering practices by asking advanced learners to apply systems thinking as they attempt to reverse-engineer a retrofitted bridge. To elaborate on mathematical practices, advanced learners could use theorems about congruent triangles to design a safe, sturdy bridge. In this particular example, the NGSS Core Disciplinary Idea of Engineering Design (MS-ETS1) is at the heart of the learning experience. Specifically, the problem illustrates NGSS ETS.1A (Defining and Delimiting Engineering Problems), and ETS1.B (Developing Possible Solutions), both important concepts taught at the middle school grades. Furthermore, students are engaged in four of the eight Science and Engineering Practices (Asking Questions and Defining Problems; Developing and Using Models; Analyzing and Interpreting Data; and Engaging in Argument from Evidence). As you can see, project-based learning experiences not only provide multiple, authentic interdisciplinary encounters with content, but they are also viable pathways to challenge scientifically and mathematically advanced learners.

Problem- and Project-Based Learning and Science Fairs

Science fair projects are a natural extension of problem- and project-based learning resulting in concepts, processes, and projects to be shared. Two fundamental differences between science fair projects and the two PBLs lie in the curricular add-on treatment and autonomous nature of science fair projects. However, science fair projects provide an avenue for students to hone their problem-solving skills and directly address NGSS Science and Engineering Practices. The integrative nature of science fairs prepare students for college, help students learn key academic content, and practice 21st-century skills such as collaboration, communication, creativity, and critical thinking. Unlike typical assignments, science fair projects do not impose limitations that result in ceiling effects, a persistent problem for gifted students. The purpose of a science fair project is similar to the purposes of both problem- and project-based learning—all three intend to teach significant content, require critical thinking, problem solving, collaboration, various forms of communication, and are organized around open-ended questions. Problem-based learning initiates inquiry through the three "What's" or "Need to Know's" (i.e., the essential content and skills), whereas science fair projects typically utilize a scientific method format to organize research. Both PBLs and science fairs allow students to take responsibility when asked to make choices, and all three have provisions for revision and reflection. With so many similarities, science fairs are considered to be avenues for extending problem- and project-based learning and promoting science talent development.

In conclusion, problem- and project-based learning experiences are both viable options for meeting the needs of scientifically and mathematically advanced students. The practical application of content, engaging key elements, and authentic, real-world experiences are excellent instructional avenues for addressing the Next Generation Science Standards. Working in tandem with the two PBLs or in isolation, science fair projects reinforce scientific and mathematical practices and provide gifted students an outlet for innovation and independent learning. All three learning experiences address 21st-century skills, augment science talent development, and contribute to scientific literacy.

Chapter 5

Assessment

Educational assessment is a complex term that encompasses a range of activities and experiences and yields information useful for placement and programming decisions as well as in monitoring student progress. Educational decisions about the learner can range along a decision-making continuum of informal to formal. Through assessment we can understand students in their contexts, determine their strengths and weaknesses, differentiate instruction, promote their awareness of learning, document their learning and change, and communicate with others connected to the students, such as parents and counselors. Examples of informal assessment include those that are found in the learning experiences (LEs) and are discussed in the first part of this chapter. Educators working with gifted and advanced learners within the NGSS also need to conduct preassessments to determine which experiences are most relevant for the typical as well as the advanced learner. Classroom instruction is the key leverage point for developing and assessing students' NGSS learning (NRC, 2014).

DOI: 10.4324/9781003238522-5

Assessment Boundaries

The NGSS often include an "Assessment Boundary" within a particular standard so that teachers will know the level of knowledge and understanding that is expected for that particular standard. In Figure 5.1, the Disciplinary Core Idea (DCI) is for grade 3, Life Science, Performance Expectation 3a. The Assessment Boundary indicates that only environmental influences on inherited traits will be assessed, not the actual mechanism of inheritance. As a result, instead of discussing meiosis and mitosis and the incidents of chromosomal damage that may cause a particular abnormality in class, the discussion will be limited to factors such as malnutrition that may cause abnormalities. However, because these boundaries are guidelines for what a teacher may present in class to the typical student, teachers may want to extend the boundary for gifted and advanced learners. Thus, in this example, the teacher may provide materials that will allow gifted students to go further in depth on the topic of heredity and study the genetic mechanism of inheritance.

Informal and Formal Assessments Relevant to Implementing Learner Experiences

The NGSS provide performance expectations for each grade, but it is important to remember that these expectations are not the curriculum to be taught but instead are the foundation for assessment tools. Using these performance expectations, assessment systems can be designed and implemented to determine mastery of curriculum content. These assessment systems include preassessments, formative or ongoing assessments, and summative assessments. As a part of the learning experience, the teacher will implement both preassessment and ongoing assessment procedures, which will vary according to the grade and lesson. These assessment procedures may include both quantitative and qualitative measures.

3-LS3 Heredity: Inheritance and Variation of Traits
Students who demonstrate understanding can:
3-LS3-a. Use evidence to support explanations that traits are inherited from parents, as well as influenced by the environment, and that organisms have variation in their inherited traits.
Assessment Boundary: Environmental influences are limited to amount and quality of nutrition, injuries, learning, and location. The genetic mechanisms of inheritance are not included.

Figure 5.1. 3-LSA Heredity: Inheritance and Variation of Traits. Adapted from *The Next Generation Science Standards: For States, By States, Volume 1, The Standards* (p. 58), by NGSS Lead States, 2013a, Washington, DC: The National Academies Press. Copyright 2013 by The National Academies Press. Adapted with permission.

Teachers of gifted students must keep in mind that high-ability students require assessment of their problem-solving and application skills. To be comprehensive, an assessment must take into consideration other information about the student's learning environment such as availability of advanced curriculum and psychosocial factors such as level of motivation. The best use of curriculum-based assessment and progress monitoring are as complements to informal measures and standardized measures that comprise a comprehensive approach to assessment.

Both formative and summative assessments based on the three-dimensional aspect of the framework of the NGSS are integral parts of classroom instruction. Hence, we next describe a variety of assessments that can be used in the classroom to assess student understanding of the NGSS.

In the learning experience (LE) examples in Chapter 3, there are varying degrees of formality associated with all aspects of assessment. For example, in the Middle School Physical Science Learning Experience, the teacher uses a brief activity to determine the students' knowledge and skills related to Newton's Third Law and their understanding of the forces in the activity, whereas in the High School Life Science Learning Experience, the teacher uses a preassessment to determine the students' readiness for the activities (see pp. 63–66 and pp. 80–82).

Observations

Ryser (2011) describes observations as a qualitative form of assessment. Observations, in particular, are relevant to the informal assessment described in the LE examples. Ryser's discussion of a system of jotting down observed behavior is presented within a broader discussion of identification for gifted programming; however, taking quick notes is also applicable to the more specific situation of a LE. As described, "When teachers observe a student exhibiting this characteristic [which offers unusual or unique responses], they jot down the student's name in the box that contains this characteristic" (Ryser, 2011, p. 45). For example, in the Middle School Genetics LE, the teacher uses observations during the activity to pinpoint any misconceptions or misunderstandings.

The validity of these notes is dependent on the teacher's ability to capture the observation at an opportune moment. In a class of 20 or more students, it is difficult to remember several different observations while searching for paper and pencil! Some teachers use flip charts, others use sticky notes or printed labels attached to a clip board. For example, the teacher may observe the students as they share their thoughts during a laboratory activity or during the process of designing an experiment and jot down notes for a formative assessment of each student. The teacher may ask herself, which child is struggling with the process? Which child is moving rapidly and accurately through the material? Whose answers show more thought and insight? (See Adams & Pierce, 2006, p. 25–26 for directions on making a flip chart.)

Observation Scales

These may be scales or checklists and are ways to note particular target behaviors. For example, the teacher may develop a checklist of group skills to use as she moves around to each group in her classroom. She may want to check for cooperation, collaboration, interdependence, accountability, honoring others' opinions, etc. This is particularly helpful in the science classroom where the emphasis is on designing and conducting investigations that require all group members to assist in the process.

Performance-Based Assessments

Performance-based assessments allow for a more quantitative approach in observing the learner's behaviors. Performance-based assessments include an evaluation component, but are typically documented through a predetermined rubric. These types of assessment are essential in science. The NGSS includes Science and Engineering Practices in each standard. *A Science Framework for K–12 Science Education* (NRC, 2012) is the blueprint upon which the NGSS are based. A major tenet of this framework is the idea that science content is melded to inquiry and discourse: Standards and performance expectations that are aligned to the framework must take into account that students cannot fully understand scientific and engineering ideas without engaging in the practices of inquiry and the discourses by which such ideas are developed and refined. At the same time, they cannot learn or show competence in practices except in the context of specific content (NRC, 2012, p. 218). Hence, engaging in scientific inquiry is expected of all students in every grade. A performance assessment is used to see if students can apply, analyze, synthesize, and/or evaluate the curriculum covered.

Rubrics. Specific criteria are used to score the task. Figure 5.2 shows a rubric that may be used for assessing students' ability to design a scientific investigation. Because there is a Form A (Do bees like diet cola?) and Form B (Do earthworms like light?), it may be used as a pre-/postassessment. Students are given the prompt and asked to design a fair test. The directions for Form A are as follows: *Explain how you would test this question: Do bees like diet cola? Be as scientific as you can as you write about your test. Write down the steps you would take to find out if bees like diet cola.* For Form B, substitute the question: *Do earthworms like light?*

Exit cards. Exit cards, sometimes called door passes or tickets to leave, are another format used to gather information

Fowler Science Process Skills Assessment
Pretest/Posttest Scoring Sheet

Name of Student _____ School _____

Score one point on student paper for each item incorporated into design. Score two points if more than one subitem is listed for a specific item.

Pre			Post
	plans to practice **SAFETY**		
	states **PROBLEM** or **QUESTION**		
	PREDICTS outcome or **HYPOTHESIZES**		
	lists more than **3 STEPS**		
	arranges steps in **SEQUENTIAL** order		
	lists **MATERIALS** needed		
	plans to **REPEAT TESTING** and tells a reason		
	other items listed by student but not on list		
	DEFINES the terms of the experiment: "attracted to," "likes," "bees," "diet cola"	**DEFINES** the terms of the experiment: "attracted to," "likes," "earthworms," "light"	
	plans to **OBSERVE**		
	plans to **MEASURE** (e.g., linear distance between bees, and/or cola, number of bees, time involved	plans to **MEASURE**: (e.g., linear distance between worms, and/or light, number of worms, time involved, amount of light)	
	plans **DATA COLLECTION:** graph or table, note taking, labels		
	states plan for **INTERPRETING DATA:** comparing data; looking for patterns in data; in terms of definitions used; in terms of previously known information		
	states plan for making **CONCLUSION BASED ON DATA** (e.g., time to notice drinks; bees may not be hungry; distances to sodas are equal; time involved for two samples is equal; temperature, light, wind, etc. are equal)	states plan for making **CONCLUSION BASED ON DATA** (e.g., time to notice light; distances to light and shade are equal; time involved for two samples is equal; temperature, wind, etc. are equal)	
	plans to **CONTROL VARIABLES** (e.g., bees not hungry; bees choose diet or regular soda; distances set equally; amounts of soda equal; number of bees tested are equal; temperature, light, wind, etc. are equal)	plans to **CONTROL VARIABLES** (e.g., worms choose dark or light; distances set equally; number of worms tested are equal; time involved is equal; temperature, wind, etc. are equal)	

Pretest score:_____ Name of rater:_____ Date:_____

Posttest score: _____ Name of rater:_____ Date:_____

Source: Fowler, M. (1990). The diet cola test. *Science Scope, 13*(4), 32–34

Figure 5.2. Assessment rubric. From "The Diet Cola Test," by M. Fowler, 1990, *Science Scope, 13*, p. 32–34. Copyright 1990 by the National Science Teachers Association. Adapted with permission.

about the day's lesson. These may take several forms, such as index cards or slips of paper with predetermined prompts. As students leave the classroom before lunch or recess or at the end of the class period, they leave their completed exit cards with the teacher. The teacher reviews the cards to determine how to adjust the lesson for the next class meeting. By reading the responses, the teacher can readily see who is still struggling with the topic and who has grasped the material. Figure 5.3 is a sample exit card for a lesson on heredity.

Peer evaluation. Students share their work with each other, either in pairs, small groups, or as a whole group. Students should have opportunities to develop the means of assessment along with the teacher. The evaluation instrument could be a checklist or answers to open-ended questions. Students could exchange papers and work in a peer editing situation or give constructive comments about each other's work. For example, students can pair up with another student outside of their own lab group to review each other's lab report and suggest improvements. Teachers will have to work with students concerning appropriate ways to give feedback before using this method of evaluation.

Self-evaluation. Have students assist in developing the assessment instrument to be used in assessing their work. As with peer evaluation, questions may be asked in a variety of formats. Some examples are: I am pleased with the way my (project, model, concept map) turned out; I thought of other ideas while working on this project; I learned something new while working on this activity. Students need to help set guidelines about the appropriate use of self-evaluation and providing thoughtful feedback. "My work was good," doesn't carry much information, but "An important lesson I learned while working on my project was_____" can provide valuable information to both student and teacher. For example, students in a teacher's first-grade science class have just completed an independent project focusing on their science unit, "The Solar System." The teacher has students work with him individually to set up three criteria that relate to their specific project. Then he has them evaluate

---

EXIT CARD
NAME:
Please provide an answer for each statement or question.
1. Explain the difference between asexual and sexual reproduction.
2. In which type of reproduction will the offspring be genetically identical to the parent?
3. Name an organism that reproduces asexually.
4. Name an organism that reproduces sexually.

Figure 5.3. Sample exit card.

---

their project on those criteria either by telling him or by writing down their responses.

Product assessment. This type of assessment is usually a rating scale or a series of open-ended questions. Teachers, parents, the child, or another designated person may fill out the scale. Some focus questions might be: To what extent does the product indicate close attention to detail? To what extent is this product beyond what a student of this age would be expected to produce? Students should be encouraged to pursue real-world problems. It follows that these projects need to be evaluated by an appropriate audience. For example, middle school students have been studying engineering design and have chosen a product to demonstrate their knowledge. Invite an engineer to assist in rating the products as a practicing professional in the field. Perhaps a classification key to the trees in the city park could be presented to and evaluated by the local garden club.

Portfolio. Portfolios are ideal ways to view a student's work over a period of time. Both students and teachers, together or separately, choose work samples to place into the portfolio. Both should have input concerning the criteria for selecting a piece for the portfolio. The portfolio may be divided into several compartments, either by unit (e.g., Weather, Force and Motion, Botany), by type of sample (e.g., lab reports, journals, quizzes, self-assessments), or by other methods. The purpose of the portfolio is multifaceted. In addition to showing growth (or lack thereof) over time, it helps develop a sense of process and serves as a way to empower both the teacher and students. For example, a student selects a lab report to add to the portfolio. The teacher and student conference over the other lab reports in the portfolio. Both can construct information and inferences about the student's learning. Both can determine strengths, areas for improvement, and growth in the process of producing a quality lab report and use that information as a guide for further learning.

Preassessments. Preassessments should be given at the beginning of a new section of study to determine who already knows some or all of the information. This information can then be used for compacting the curriculum for those who already show mastery. For example, the teacher is beginning a study of force and motion in her honors physics class. She gives the students a preassessment covering the material she intends to present. Students who know major portions of the material already need to have alternate assignments rather than being required to complete work they have already mastered.

Rating scales. Rating scales can be used for peer evaluation, self-evaluation, product assessment, and other types of assessment. Statements are listed and the respondent indicates the extent of agreement with the statements. The scale may be of any length, but is usually three, four, five, or ten units. For example, a three-unit scale may be "agree, neutral, disagree." A five-unit scale may be "strongly agree, agree, neutral, disagree, and strongly disagree." Other scales may look like this:

1	2	3	4	5	6	7	8	9	10

strongly disagree   strongly agree

Suppose the teacher has just finished a unit on "The Night Sky" in his third-grade science class. He could construct a three-unit scale, using the previously suggested words; pictures of a thumb up, a thumb sideways, and a thumb down; or a smiley face, a straight face, and a frowny face. Students could circle their reactions to statements such as, "I remembered to draw the moon every night" or "I looked for patterns in my pictures of the moon."

Teacher–student and/or student–student conferences.
Conferences allow both parties to engage in a dialogue about a
particular work or assignment. For example, a teacher–student
conference about a research paper on an endangered animal may
involve the student explaining why she chose the topic, why a
particular choice of words was used, or any personal meaning
that may not be apparent.

Notebooks. Science notebooks are generally more struc-
tured, with a table of contents and numbered pages. Observations;
questions; laboratory reports, including hypotheses, procedures,
data tables, analyses, conclusions; diagrams and drawings; concept maps; and graphic organizers are among their many uses.
Science notebooks may be used as a dialogue between teacher and
student on any topic. Students have the opportunity to express
their own ideas and opinions. At the same time, teachers can gain
valuable insight into the students' interests, writing abilities, and
thought processes, as well as identify any misconceptions about a
particular science topic (Gilbert & Kotelman, 2005; Marcarelli,
2010).

Assessment of Science Achievement

Achievement tests were developed to measure what students have learned and are built on the assumption that the test
measures what students have learned *recently*. This is a reason-
able assumption for the typical student, yet not necessarily for
the advanced learner. In fact, advanced learners are oftentimes
considered advanced because of their very high performance
on grade-based achievement tests, but they may have learned
the material well before entering the grade level. Furthermore,
demonstrated high performance may be regarded as indicative
of mastery of skills measured by the test, which is generally just
grade-level mastery. Traditionally, science achievement measures have consisted of multiple-choice, short-answer, true or
false, and other formats that assess mainly grade-level factual
knowledge. Grade-level achievement tests have been serving
schools, teachers, and students for decades. However, these tests

were not designed to measure the kinds of learning that should occur when curriculum is based on the NGSS (NRC, 2014). Recall that the NGSS focus on three Dimensions: Disciplinary Core Ideas, Crosscutting Concepts, and Science and Engineering Practices. Hence, measuring this three-dimensional science learning requires "assessment tasks that examine students' performance of scientific and engineering practices in the context of disciplinary core ideas and crosscutting concepts" (NRC, 2014, p.86). Indeed, the performance expectations, which indicate what students should be able to know and do, articulate what should be assessed at a particular grade level. Students will be required to apply knowledge and practices, not just know factual information. With the focus of the NGSS on deep knowledge and 21st-century skills, the traditional format for testing science achievement—having students respond to multiple-choice questions at one sitting—will have to change:

> It will not be feasible to assess all of the performance expectations for a given grade level during a single assessment occasion. Students will need multiple—and varied—assessment opportunities to demonstrate their competence on the performance expectations for a given grade level. (NRC, 2014, p. 2)

Additionally, The NGSS "require that assessment tasks be designed so that they can accurately locate students along a sequence of progressively more complex understandings of a core idea and successively more sophisticated applications of practices and crosscutting concepts" (NRC, 2014, p. 87). Although there is mention of assessing students along a continuum that may be anchored by the terms "basic," "proficient," and "advanced" (NRC, 2014, p. 85), there is nothing in the *Developing Assessments for the Next Generation Science Standards* report addressing out-of-level testing or gifted learners. It remains to be seen whether the needs of gifted and advanced learners will inform the assessments as these are being constructed (e.g., will these assessments have a high enough ceiling?). Moreover, Duncan and Rivet (2013) acknowledge that determining the levels of student knowledge and reasoning across a learning progression is quite challenging:

> Students at the high end of a LP tend to demonstrate reasoning that is robust and consistently applied to diverse assessment tasks. Students at the beginning of a LP typically perform poorly across items. Yet, at intermediate levels of a LP, students' levels of understanding may vary from item to item because their developing knowledge is not robust enough to be consistently applied to diverse situations and phenomena. (p. 397)

Hence, assessments will need to be mindful of both context and features of the items as well as "how these are correlated with the intermediary steps in a LP" (Duncan & Rivet, 2013, p. 397).

Computer-Based Assessments

We are now in the digital age, and 21st-century assessments are clearly a reflection of this reality, although 21st-century language is being replaced by reference to the "next generation." The first computer-based test to be discussed is the Measure of Academic Progress (MAP). MAP is a computer-adaptive test that was released in 2005, and thus is a relatively "old" test compared to the three tests scheduled for release in the fall of 2014: Smarter Balanced, PARCC, and Act Aspire. Currently, Smarter Balanced and PARCC do not have science assessments; thus we will only discuss MAP and ACT Aspire.

Measure of Academic Progress (MAP). The Measure of Academic Progress (MAP) test, developed by the Northwest Evaluation Association (NWEA), was one of the first computer-adaptive tests to be used on a large-scale basis. MAP assessments are similar to traditional assessments in that they correspond to state-based standards and assess in the traditional areas of reading, math, and language arts. Recently, the science tests, aligned to the former National Science Education Standards, were designed and can be used in grades 3–10. They have not yet been realigned to the NGSS. The MAP assessments have at least two advantages over the more traditional paper-and-pencil measures of achievement. First, the tests are adaptive. In other words, they are built upon a statistical process known as Item Response Theory (IRT). The item difficulty adjusts to the student's response. Second, the tests were designed to be used up to four times during an academic year. Finally, although the MAP tests were developed relatively recently (2005), NWEA was incorporated in 1977 and has a very strong reputation among educators.

ACT Aspire. ACT Aspire, developed by ACT, Inc., a company known for its college entrance exam, the ACT, is a digital longitudinal assessment program that tracks progress from elementary grades through high school with a focus on college and career readiness. ACT Aspire offers tests in the same areas provided by ACT's current suite of assessments in English, math, reading, and science, beginning with grade 2 and continuing through high school. The Classroom Quizzes science component of the ACT Aspire Formative Assessment Solution is designed with alignment to the Next Generation Science Standards and to ACT's College and Career Readiness Benchmarks. For more information on the ACT Aspire system, visit http://www.dis coveractaspire.org/.

Once schools adopt the NGSS, it is obvious that new forms of assessment will be needed to evaluate students' learning at the classroom, district, and state levels. The traditional formats used by states to monitor progress will not suffice. Assessment of the NGSS begins at the classroom level using formats that allow documentation of progress toward meeting the performance expectations, and these assessments must have high ceilings to accommodate gifted and advanced learners. Professional development and adequate support for teachers are vital for the appropriate implementation of new curriculum and assessments that

1. Do your classroom assessments have multiple and varied opportunities to demonstrate NGSS blended knowledge?
2. Do your classroom assessments have multiple component tasks (sets of interrelated questions) for a given NGSS performance expectation?
3. Do your classroom assessments focus on or highlight a smaller set of most important gatekeeper concepts?
4. Have you thought about or tried scaffolds or hints in assessment tasks to help scaffold students' ability to generate valuable information about NGSS blended knowledge?
5. Are you careful to score/provide feedback on DCIs, practices, and blended knowledge products?

Figure 5.4. Checklist of recommendations for NGSS assesments.

address the NGSS. Figure 5.4 provides a checklist that teachers and administrators can use to determine if classroom assessments are appropriate for assessing students' progress toward mastery of the NGSS.

In presenting this information on assessments, we are suggesting that students and teachers are likely to benefit from a comprehensive approach to assessment. Preassessment has an important place at the table. So, too, does informal assessment of the learner's understanding of the material presented in the LE. Formal assessment procedures are catching up with the ever-changing world of computers and now reflect the impact of technology in terms of delivery of items as well as with regard to reporting results.

Today's educators have a variety of options. There are many tests that have been developed by large-scale testing companies that can provide valuable insight and information into an individual learner's needs as well as an understanding of that learner's needs within the context of a class of students. These assessments are valid and reliable; however, the usefulness rests with the educator who must connect the information to the learner's needs.

Chapter 6

Management Strategies

The NGSS recommend that K–12 students continuously engage in science and engineering practices (NGSS Lead States, 2013a). With this recommendation come questions about how to manage such a classroom. Managing any science classroom, especially one that is student centered and full of activity, can be intimidating even for the most seasoned teachers. Without proper classroom management strategies, teachers are frazzled by the end of the day and their classrooms are disheveled. More importantly, teachers will be reluctant to provide these types of learning activities for their students if classroom control is difficult to maintain (Frazier & Sterling, 2005). This chapter will provide readers with common strategies used in science classrooms that will help make science learning enjoyable for both the students and the teacher.

In order to establish an effective and supportive classroom, teachers need to consider both the classroom culture they create with their students as well as the design and structure of their lessons. Although lesson design may not seem to be related to classroom management, lessons that engage students tend to lead to less disruptions and behavior issues.

DOI: 10.4324/9781003238522-6

Classroom Culture

To properly engage students in the practices of science and engineering, students need a classroom culture where thinking, problem solving, and collaboration are promoted. Students must also feel they are free to take risks in their learning without a fear of failure. Conversely, in classrooms where the teacher is the sole decision maker and the supplier of all knowledge, students often depend on the teacher's thinking and they struggle to solve problems individually or in groups (Manning et. al., 2009). Students acculturated to didactic, teacher-centric classrooms find engaging in cooperative learning groups to be difficult because they prefer the teacher provide all the answers rather than depend on themselves and other students to solve the problem. In essence, they are reluctant to trust anyone but the teacher to provide the correct answer.

Before seeking to change teaching practices to a more student-centered classroom, Manning and colleagues (2009) recommend teachers begin working with students in the area of classroom management. Teachers should first work with students to establish an autonomous classroom environment that is characterized by mutual respect between the teacher and students and among students. Specifically, Manning and colleagues (2009) suggest that teachers involve students in rule setting and conflict resolution. For example, when a conflict arises in the classroom, the teacher will engage students in a discussion about the conflict as opposed to silencing the students and moving on with the task at hand. Through this process, students will learn to compromise and negotiate in reaching a mutual solution while demonstrating respect for all parties involved. The researchers maintain this type of conflict resolution will provide real-world examples of problem solving and collaboration that are desired in the science classroom. In this way, students will extend collaboration and problem-solving skills learned in classroom management to working with science content.

The classroom culture should also be one in which the expectations are high and the communication is positive. Precise academic and behavioral expectations need to be apparent to all students with a clear understanding of the consequences if either area is violated (McIntosh, 2009; Sterling, 2009). In relaying expectations and consequences, teachers should model positive communication. As McIntosh (2009) reiterated, one positive word has more power to motivate a child than 20 negative words. Through teacher modeling of positive communication to students, students should develop the habit of positive communication with their classroom peers, thereby strengthening the classroom community (Wolfgang, 2009).

To encourage a classroom culture of student-centered learning and create a strong classroom community, Wolfgang (2009) recommends teachers utilize team-building activities. These activities facilitate trust and risk-taking behaviors, serving to increase cooperation and respect among classroom peers. Table 6.1 offers examples of such activities designed to strengthen the classroom community. These activities are great first-day exercises or supplements to be used before students are released from school for a holiday.

Organized and Engaging Lessons

Another means to establishing and maintaining classroom control is through organized and engaging lesson plans (McIntosh, 2009; Poon, Tan, & Tan, 2009; Romano, 2012). A guaranteed way to lose student interest is through a boring lesson plan or being unprepared. Many teachers have committed the unpardonable sin of entering the classroom not fully prepared. In these instances, student learning is scarce, and by the end of the day these teachers are ready to give up teaching. The old saying of "monitor and adjust" is not an excuse for a lack of preparedness. No amount of monitoring or adjusting can make up for an ill-prepared lesson.

Table 6.1

Team-Building Activities

Team-Building Activity	Description	Resources
Human Checkers	In a game of checkers, two students take the role of players and the remainder of the class mimics the game pieces. Players and student game pieces work together to defeat the opposing team.	http://www.leadership-with-you.com/fun-team-building-exercises.html
Leader vs. Follower	Students demonstrate team-building skills while learning to be effective leaders and followers.	https://www.ffa.org/documents/learn/LOG.FFA_Leadership.pdf
Helium Stick	In this game, students work together in small or medium-sized groups to lower a stick to the ground. The trick is that each person's fingers must be in contact with the helium stick at all times.	http://wilderdom.com/games/descriptions/HeliumStick.html
Amoeba Race	This activity requires students to create the parts of an amoeba (protoplasm, cell wall, and nucleus) and mimic its movement. The amoeba will eventually split and the two amoebas can race.	http://wilderdom.com/games/descriptions/AmoebaRace.html
Magic Carpet Ride	With everyone standing on a full-sized bed sheet, participants work together to turn the sheet over without anyone touching the floor or ground.	http://cchealth.org/tobacco/pdf/activities.pdf

Advanced planning is the key to a smooth laboratory or hands-on activity in the science classroom. Having all materials organized and available enables transitions from instructions to activities to move much more efficiently. Wolfgang (2009) advises teachers to use plastic storage containers to hold items students may need for an activity. You could have containers for each group or containers with the supplies located in a central location for easy pick-up. Romano (2012) and Sterling (2009) also caution teachers to make sure the experiment works as intended and that it takes the amount of time that is expected. Time spent in planning saves time during instruction and alleviates confusion during the class period.

To keep students engaged and to address the needs of all learners, planning of student grouping is essential in the management of a classroom. As noted by previous studies, within-class flexible grouping options have resulted in higher achievement among both advanced and middle-achieving learners (Tieso, 2005). Student grouping should not be static but instead based on the individual lesson and the desired learning outcome. The profile of the groups, as determined through preassessment, could be formed according to student readiness, interest, or learning style/ multiple intelligences. Or in the case of a problem-based unit, student groups might be based on their viewpoint concerning a particular topic. For example, if students were investigating the pros and cons of nuclear energy (Center for Gifted Education, 1999), students with similar opinions could be the determinant for the grouping.

To improve the structure and focus of student thinking and to ensure all students participate in and contribute to the task at hand, individual students need definite roles or jobs within their groups (Krajcik, Czerniak, & Berger, 2003; Michaels, Shouse, & Schweingruber 2008). Krajcik and colleagues described three types of roles: interpersonal roles, managerial roles, and cognitive roles. Students assigned to an interpersonal role are involved in making sure the group works well together. These roles may include facilitators, encouragers, and noise monitors. Students assigned to a managerial role include the readers, timekeepers, recorders, and runners (material managers). Finally, those assigned to cognitive roles are those who require a specific type of thinking or knowledge, such as those used in problem- or project-based scenarios. As in our previous example, if students are studying the effects of nuclear energy on the environment, students might take the role of various stakeholders such as a city council member, nuclear scientist, environmentalist, and community member (Center for Gifted Education, 1999). Krajcik et al. (2003) recommend the use of cognitive roles for upper elementary to secondary students. They claimed older students tend to become bored with the nonchallenging tasks of the interpersonal and managerial roles. In the same way, advanced learners may find cognitive roles more appealing and stimulating than interpersonal and managerial roles.

To address strengths and weaknesses of individual student performance and encourage high expectations, Wolfgang (2009) recommends students complete a self-assessment of their role in the group activity. Table 6.2 provides an example of the self-assessment tool used by Wolfgang.

Classroom management can be enhanced by a lesson that is both intriguing and engaging. The best teachers often use a hook to garner students' attention at the beginning of the class. This is easily done in science class by providing the explanation after the experiment, in a sense, *putting the cart before the horse*. By doing the experiment before the explanation, teachers give students a reason to pay attention. For example, in learning about chemical reactions, teachers can do a demonstration or better yet, allow students to perform a simple experiment to guide them in recognizing the indicators of a chemical reaction. A simple experiment could be blowing up a balloon by mixing baking soda and vinegar and allowing students to observe and reflect on the changes. After the experiment, students would describe the changes that occurred and the teacher could help students fill in the missing informational pieces that will lead to conceptual understanding.

Michaels, Shouse, and Schweingruber (2008) maintain that sequencing is necessary for meaningful instruction. They recommend teachers use the "just in time" (p. 129) approach by providing students the knowledge and skills at the points they are needed in an investigation, where they will be applied to develop new ideas. For example, in continuing our study of nuclear

Table 6.2

Self-Assessment of Role Assignments

Number	Role	Duty	Student name	These were my strengths while performing my role.	These are things I should work on next time while performing this role.
1	Materials manager	Gather, maintain, and return materials.			
2	Recorder	Record notes and data from all tests.			
3	Safety Engineer	Monitor and promote safety; report safety breaches.			
4	Intelligence officer	Read instructions; gather additional information.			
5	Test pilot	Perform tests.			

energy, students are required at the end of the unit to make a recommendation for or against using nuclear power for energy usage (Center for Gifted Education, 1999). Students would not be prepared to make this recommendation at the onset of the unit. After the problem is introduced, students are taken through various scenarios where they learn how energy is produced at nuclear power plants and about the safety guidelines that must be used. At strategic points throughout the unit, students are introduced to complex concepts that allow them to gather evidence to support their eventual conclusion. In addition to "just in time" teaching, teachers should be prepared to adapt instruction and instructional materials as needed. If it is clear that instruction is not going as intended, it is important for teachers to redirect student learning toward the learning outcomes.

Just as a strong hook at the beginning of the lesson is important, so is the lesson closure. Sterling (2009) suggested teachers leave enough time at the end of the class for clean-up and reflection of the learning activity. Clean-up is important so the teacher will not be rushed during the transition to another class or another subject while reflection gives students opportunities to conceptualize their learning.

Teachers may want to use a lesson plan template or instructional model to help develop an engaging lesson plan. The 5E Learning Cycle is a commonly used instructional model that was modified and developed by the Biological Sciences Curriculum Study (BSCS) in the 1980's (Bybee et al., 2006). This instructional model consists of five phases: engagement, exploration, elaboration, and evaluation designed to effectively engage students in learning and predicate their conceptual understanding of the topic being explored. The effectiveness of the instructional model was reported by Bybee et al. (2006). They reported positive results from using the 5E Learning Cycle on students' science knowledge and skills and their attitudes and interest in science; however, they suggested more studies be conducted to adequately measure the effectiveness of the instructional model. Nevertheless, an instructional model that is designed to actively engage students in learning is a tool worth using in the management of a classroom. Table 6.3 provides a summary of each phase of the instructional model.

The importance of managing a science classroom cannot be overemphasized. As recommended by the NGSS, students need to be actively involved in the practices of scientists and engineers. This is difficult to do if classroom control is not maintained. The strategies recommended in this chapter to help with classroom management are tools designed to effectively engage students in active learning. Classroom management issues can be addressed and establishing a classroom culture where students are encouraged to think and develop solutions to problems can encourage active learning. The classroom culture can be improved through (a) encouraging collaboration among students in cooperative learning groups, (b) demonstrating mutual respect between the teacher and fellow peers, and (c) building camaraderie among classmates in establishing a community of learners. Organized and engaging lessons designed for student-centered learning also should be in a teacher's classroom management toolbox. Students will more readily respond to a well-prepared lesson that captures their attention, makes them think, and allows them to generate solutions on their own or within their cooperating group (Poon, Tan, & Tan, 2009). To ensure lessons are organized and engaging, teachers should plan for lessons and materials needed in advance, assign students roles within the lesson structure, hook the students through a captivating introduction, provide students with "just in time" knowledge and skills, and allow an adequate amount of time for closure.

All of the strategies presented in this chapter necessitate planning on the teacher's part. In particular, the teacher needs to plan how to establish a community of learners before the first day of school. The teacher needs to plan every detail of the lesson and activity to best promote student engagement and intrigue. The teacher also needs to prepare materials in such a way to minimize disturbances during transitions. In conclusion, time that is well spent in planning will eliminate many of the typical classroom management problems seen during class time.

Table 6.3

Summary of the BSCS 5E Instructional Model

Phase	Summary
Engagement	The teacher or a curriculum task accesses the learners' prior knowledge and helps them become engaged in a new concept through the use of short activities that promote curiosity and elicit prior knowledge. The activity should make connections between past and present learning experiences, expose prior conceptions, and organize students' thinking toward the learning outcomes of current activities.
Exploration	Exploration experiences provide students with a common base of activities within which current concepts (i.e., misconceptions), processes, and skills are identified and conceptual change is facilitated. Learners may complete lab activities that help them use prior knowledge to generate new ideas, explore questions and possibilities, and design and conduct a preliminary investigation.
Explanation	The explanation phase focuses students' attention on a particular aspect of their engagement and exploration experiences and provides opportunities to demonstrate their conceptual understanding, process skills, or behaviors. This phase also provides opportunities for teachers to directly introduce a concept, process, or skill. Learners explain their understanding of the concept. An explanation from the teacher or the curriculum may guide them toward a deeper understanding, which is a critical part of this phase.
Elaboration	Teachers challenge and extend students' conceptual understanding and skills. Through new experiences, the students develop deeper and broader understanding, more information, and adequate skills. Students apply their understanding of the concept by conducting additional activities.
Evaluation	The evaluation phase encourages students to assess their understanding and abilities and provides opportunities for teachers to evaluate student progress toward achieving the educational objectives.

Chapter 7

Professional Development

The NGSS were designed to make real and valuable changes in science education. Unfortunately, many teachers are not prepared for the instructional changes suggested by the NGSS. Numerous teachers, particularly elementary teachers, lack the science knowledge and conceptual understanding needed to facilitate investigations and discussions, which can be the most challenging factor when addressing the needs of gifted science learners (Coates, 2006). Elementary teachers, in particular, have had very little science preparation in their undergraduate degrees, and if they did have science courses, many of them were instructed through lecture methods, giving teachers no real insights to how science should be taught (Fulp, 2002; Michaels, Shouse, & Schweingruber, 2008). In contrast, secondary teachers have experienced multiple science content courses but may lack the pedagogical knowledge to teach science effectively as recommended by the NGSS (Roehrig & Luft, 2006). Too often in secondary courses the content is vast, and teachers are pushing to get through the standards so students will be prepared for end-of-course or Advanced Placement examinations (Schwartz, Sadler, Sonnert, & Tai, 2009). This type of broad content attain-

DOI: 10.4324/9781003238522-7

ment limits the in-depth study of a topic as recommended by the NGSS.

The question remains: How will we prepare teachers to make the needed changes in their classroom to effectively implement the NGSS and sufficiently challenge advanced learners? Unless teachers are provided with the proper training and support, the NGSS could face the same outcome as previous science standards—they will probably be placed on a bookshelf and collect dust with other standards, or they will be used as a checklist to mark off topics covered and not used as a guide to in-depth study. The risk in the implementation of the NGSS lies in the lack of teacher training and support, especially for advanced learners. There are fewer standards, which could lead some to the false belief that they need to spend less time in science and superficially cover the highlighted topics. If this is the case, in-depth learning that leads to conceptual understanding will be limited in the classroom environment.

One key to effectively implementing the NGSS is professional development and support for teachers. The professional development must be empirically validated, be able to promote and extend effective curricula and instructional models, and be intensive, sustained, content focused, well defined, and strongly implemented (Yoon, Duncan, Lee, & Shapley, 2008). Additionally, to promote extended or accelerated learning opportunities for gifted students using the NGSS as a springboard, teachers need ongoing professional development and support in quality learning endeavors, such as inquiry- and problem-based learning (Little & Paul, 2011; Marshall & Horton, 2011; Robinson, Dailey, Hughes, & Cotabish, 2014).

Professional Development: Critical Elements

The National Research Council (NRC) recommended that science professional development offer teachers sustained follow-up support, learning using classroom-specific curriculum, information on how students learn science, and models of how to teach science (Duschl, Schweingruber, & Shouse, 2007). It should focus on the curriculum teachers will teach in their classroom and mirror the instruction that will occur (Duschl et al., 2007). To make the necessary adjustments in teacher instruction to effectively teach science as suggested by the NGSS, researchers recommended the following critical elements of professional development be addressed: extended contact time, embedded or follow-up support, collective participation of teachers, curriculum- and content-specific training, and active learning (Anderson, 2007; Buczynski & Hansen, 2010; Cotabish, Dailey, Hughes, & Robinson, 2011; Cotabish, Dailey, Robinson, & Hughes, 2013; Desimone, Porter, Garet, Yoon, & Birman, 2002; Lumpe, Czerniak, Haney, & Beltyukova, 2012; Penuel, Gallagher, & Moorthy, 2011).

Extended Contact Time

Darling-Hammond and Richardson (2009) maintained that one-shot attempts at professional development or those that required teachers to make changes without support have not been effective in changing teacher instruction. As a result, they recommended sustained teacher training and support over multiple days or weeks. With this in mind, the Council of Chief State School Officers (CCSSO, 2008) reported the effects of professional development programs in math and science increased with a minimum of 45 hours of contact time each year, including summer institutes and in-classroom support. In relation to inquiry-based instruction, Supovitz and Turner (2000) found teachers needed a minimum of 40 to 79 hours of science-specific professional development to establish an investigative classroom. In another study, Gerard, Varma, Corliss, and Linn (2011) found that 67% of teachers who participated in a constructivist-oriented professional development program for more than a year increased students' frequency of inquiry-based learning opportunities. In contrast, 67% of teachers who had participated in the professional development for less than a year tended to revert back to a direct instructional approach.

Unfortunately, teachers are not lined up at the door eagerly volunteering to participate in extended professional development programs. So, with this knowledge, how can we encourage teachers to participate in lengthy professional development programs? Offering teachers stipends for summer institutes helps, as does providing them with support during the school year to implement these new changes. Cotabish, Dailey, Hughes, and Robinson (2011) found that a professional development model that utilized stipends for summer institutes and provided job-embedded support resulted in positive changes in teachers' instructional practices. Additionally, these changes in teacher instruction resulted in increased science learning among both gifted and general education students (Cotabish, Dailey, Robinson, & Hughes, 2013; Robinson et al., 2014).

Embedded Support

One method of providing sustained and embedded support involved a professional development model incorporating peer coaching or mentoring. The utilization of a peer coach or mentor enabled teachers to practice authentic instructional scenarios in the context of their own classroom with the help of another peer or expert. Dailey and Robinson (2013) found teachers experienced increased self-efficacy for teaching science and improved science process skills after participation in a program that provided 60 hours of summer institutes with 60 hours of peer coaching support across 2 years. More remarkably, they found these changes were sustained one year after teachers completed the program.

Peer coaching or mentoring in-service teachers can occur in a variety of ways. Designed to increase student learning and talent development in the STEM disciplines, one study used an "expert" peer coach that visited both gifted pull-out classrooms and general education classrooms periodically throughout the school year (Cotabish et al., 2011; Cotabish et al., 2013; Dailey & Robinson, 2013; Robinson et al., 2014). The "expert" was a former high school science teacher and gifted educator who had experience in providing professional development to teachers. During the classroom visits, the peer coach would often model teaching strategies, act as an instructional facilitator, and provide and set up materials for the science activity. Other studies have used university professors, scientists, or graduate students to take the role of the "expert" peer coach (Appleton, 2008).

Regrettably, schools do not always have access to "expert" peer coaches. Schools could consider using their science specialists or high school science teachers as peer coaches, but this is often difficult because of scheduling conflicts. Another option, as described by Ackland (1991), uses peer teachers as reciprocal peer coaches. Reciprocal peer coaches alternately observe, provide feedback, and engage with each other in collaborative discussions about their teaching. Expert peer coaching is recommended when implementing a new instructional strategy as in the NGSS (Swafford, 1998); however, this may not be possible due to financial and scheduling demands. In these instances, reciprocal peer coaching would be the method of choice.

Collective Participation

Implementation of a new innovation such as NGSS typically will be more successful if multiple teachers from the same school participate in the professional development. The Council of Chief State School Officers (2008) found the most effective programs involved a collective participation of teachers from either the same grade level, department, or school. In schools where multiple teachers participated in the professional development program, Buczynski and Hansen (2010) and Desimone, Porter, Garet, Yoon, and Birman (2002) reported the effect of the professional development program on instructional practices was strengthened. This finding is typical and explainable.

When teachers are given opportunities to reflect on their learning and collaborate with their colleagues, they are more likely to respond positively to the professional development experience and in turn, implement the strategies learned in their classrooms (Wojnowski & Pea, 2014). Additionally, including teachers of gifted programs in the collaborative groups gives voice to advanced learners. Through participation, teachers of gifted students could offer suggestions on how to challenge advanced learners by differentiating the NGSS.

Content-Specific Professional Development

It is important for the content of the professional development to be meaningful to the participating teachers. Many experts recommend that teachers be engaged in the content and practices from their grade-specific standards (Duschl et al., 2007). In a recent study, Dailey and Robinson (2013) reported that teachers expressed greater satisfaction with the professional development program when they received specific training on their classroom-based curricula. This is especially true for elementary schools, where science content knowledge may be limited and learning an abundance of new material can be overwhelming for teachers. In another study, Penuel, Gallagher, and Moorthy (2011) compared three professional development programs. The two that provided teachers with explicit instructions on their specific models of teaching resulted in the greatest gain in student achievement. Others, such as Kennedy (1998), maintained the content of the professional development is more important than extended contact time. Little improvement will be made in the classroom if the content is not relevant to the teachers' instructional assignment.

Active Learning

Finally, professional development programs that provide teachers with opportunities to engage in active learning are usually more successful at improving instructional practices (CCSSO, 2008; Desimone et al., 2002; Little & Paul, 2011). Especially with regard to science learning, teachers need to be involved in the "doing" of science as opposed to being a passive listener of science. Dailey and Robinson (2013) found that teachers preferred to take the role of the student as the professional development instructor guided them through their classroom-specific unit lessons. They commented that being able to complete the activities and experiments that would be used in their own classrooms provided them with the perspective of the student and enhanced their own understanding. Other successful professional development programs utilized active learning by engaging participant teachers in modeling instruction, planning lessons, coaching teachers, developing assessments, and observing and collaborating with other teachers (CCSSO, 2008).

If we want teachers to effectively implement the NGSS and enact real change in their classrooms, they must have the opportunity to develop the knowledge and skills to make this possible. The litmus test to determining the support of these new standards is in the amount of resources national, state, and local policy makers, as well as school districts, are willing to provide to ensure effective implementation. Effective implementation will require that teachers be provided with opportunities to increase their knowledge and skills through quality professional development programs that engage teachers in sustained and embedded support, collective participation, and content-specific and active learning.

References

Achieve, Inc. (2014a). *Next Generation Science Standards.* Washington, DC: Author.

Achieve, Inc. (2014b). *Three dimensions.* Retrieved from http://www.nextgenscience.org/three-dimensions

Ackland, R. (1991). A review of the peer coaching literature. *Journal of Staff Development, 12,* 22–26.

Adams, C. M., Cotabish, A., & Ricci, M. C. (2014). *Using the Next Generation Science Standards with gifted and advanced learners.* Waco, TX: Prufrock Press.

Adams, C. M., & Pierce, R. L. (2006). *Differentiating instruction: A practical guide to tiered lessons in the elementary grades.* Waco, TX: Prufrock Press.

Akinoglu, O., & Tandogan, R. O. (2007). The effects of Problem-Based Active Learning in science education on students' academic achievement, attitude and concept learning. *Eurasia Journal of Mathematics, Science, & Technical Education, 3*(1), 71–81.

Anderson, R. D. (2007). Inquiry as an organizing theme for science curricula. In S. K. Abell & N. G. Lederman (Eds.),

Handbook of research on science education (pp. 807–830). Nahway, NJ: Lawrence Erlbaum Associates.

Appleton, K. (2008). Developing science pedagogical content knowledge through mentoring elementary teachers. *Journal of Science Teacher Education, 19*, 523–545.

Benbow, C. P., & Stanley, J. C. (1996). Inequity in equity: How current educational equity policies place able students at risk. *Psychology, Public Policy, and Law, 2*, 249–293.

Blunier, T., Daugherty, M. K., & Morford, L. (2009). *A collection of engineering design problems*. Fayetteville: University of Arkansas. Retrieved from http://cied.uark.edu/2009_UA_PROBLEM_SOLVING.pdf

Buck Institute for Education. (n.d.). *What is project-based learning (PBL)?* Retrieved from http://bie.org/about/what_pbl

Buczynski, S., & Hansen, C. B. (2010). Impact of professional development on teacher practice: Uncovering connections. *Teacher and Teacher Education, 26*, 599–607.

Bybee, R. W., Taylor, J. A., Gardner, A., Van Scotter, P., Powell, J. C., Westbrook, A., & Landes, N. (2006). *The BSCS 5E instructional model: Origins and effectiveness*. Retrieved from http://bscs.org/sites/default/files/_legacy/BSCS_5E_Instructional_Model-Full_Report.pdf

Callahan, C. M., & Kyburg, R. M. (2005). Talented and gifted youth. In D. L. Dubois, & M. J. Karcher (Eds.), *Handbook of youth mentoring* (pp. 424–439). Thousand Oaks, CA: Sage.

Center for Gifted Education. (1999). *Nuclear energy: Friend or foe. Examining nuclear power issues from a systems perspective*. Dubuque, IA: Kendall Hunt.

Chapman, C. (2009). A smoother acceleration. *The Science Teacher, 76*(3), 42–45.

Chapin, S. H., O'Connor, C., & Anderson, N. C. (2009). *Classroom discussions: Using math talk to help students learn*. Sausalito, CA: Math Solutions.

Cheuk, T. (2012). *Relationships and convergences found in Common Core State Standards in Mathematics, Common Core State Standards in ELA/Literacy, and A Framework for K–12 Science*

Education. Arlington, VA: National Science Teachers Association. Retrieved from http://www.nsta.org/about/ standardsupdate/resources/VennDiagram–CommonCore– Framework.pdf

Coates, D. (2006). 'Science is not my thing': Primary teachers' concerns about challenging gifted pupils. *Education, 34,* 49–64.

Colangelo, N., Assouline, S. G., & Gross, M. U. M. (2004). *A nation deceived: How schools hold back America's brightest students.* Iowa City: The University of Iowa, The Connie Belin & Jacqueline N. Blank International Center for Gifted Education and Talent Development.

Colley, K. (2008, November). Project-based science instruction: A primer. *The Science Teacher. 75*(3), 23–27.

Consortium for Policy Research in Education. (2009). *Learning progressions in science: An evidence–based approach to reform.* Retrieved from: http://www.cpre.org/images/stories/cpre_ pdfs/lp_science_rr63.pdf

Cooney, T. M., Escalada, L. T., Unruh, R. D. (2008). PRISMS (Physics Resources and Instructional Strategies for Motivating Students) Plus CD version. Cedar Falls: University of Northern Iowa Physics Department.

Cotabish, A., Dailey, D., Coxon, S., Adams, C., & Miller, R. (2014). The Next Generation Science Standards and high ability learners. *Teaching for High Potential, Winter 2014* (1), 16–18.

Cotabish, A., Dailey, D., Hughes, A., & Robinson, A. (2011). The effects of a STEM professional development intervention on elementary teachers' science process skills. *Research in the Schools, 18*(2), 16–25.

Cotabish, A., Dailey, D., Robinson, A., & Hughes, A. (2013). The effects of a STEM intervention on elementary students' science knowledge and skills. *School Science and Mathematics, 113*(5), 215–226.

Council of Chief State School Officers. (2008). *Does teacher professional development have effects on teaching and learning?* (Grant No.

REC 0438359). Retrieved from http://www.ccsso.org/proj ects/improving_evaluation_of_professional_development

Dailey, D. (2014). *What can we learn from plants and animals?* Unpublished lesson.

Dailey, D. & Robinson, A. (2013). *The effect of implementing a STEM professional development intervention on elementary teachers.* Retrieved from ProQuest (UMI 3587609).

Darling-Hammond, L., & Richardson, N. (2009). Teacher learning: What matters? *Educational Leadership, 66*(5), 46–53.

Desimone, L. M., Porter, A. C., Garet, M. S., Yoon, K. S., Birman, B. F. (2002). Effects of professional development on teachers' instruction: Results from a three-year longitudinal study. *Educational Evaluation and Policy Analysis, 24*(2), 81–112.

Dewey, J. (1938). *Experience and education.* New York, NY: Macmillan.

Duncan, R. G., & Rivet, A. E. (2013). Science learning progressions. Science, *339* (6118) 396–397.

Duschl, R., Schweingruber, H. A., & Shouse, A. (2007). *Taking science to school: Learning and teaching science in grades K–8.* Washington, DC: The National Academies Press.

Finkle, S. L., & Torp, L. L. (1995). Introductory documents. Aurora, IL: Center for Problem-Based Learning.

Flinn Scientific, Inc. (2006). *Plotting trends: A periodic table activity.* Batavia, IL: Author. Retrieved from http://cationclub.files. wordpress.com/2013/10/cf10480.pdf

Fowler, M. (1990). The diet cola test. *Science Scope, 13,* 32–34.

Frazier, W. M., & Sterling, D. R. (2005). What should my science classroom rules be and how can I get my students to follow them? *Clearing House, 79*(1), 31–35.

Fulp, S. L. (2002). *Status of elementary school science teaching.* Retrieved from http://2000survey.horizon–research.com/ reports/elem_science/elem_science.pdf

Gallagher, S. A., Sher, B. T., Stepien, W. J., & Workman, D. (1995). Implementing problem–based learning. *School Science and Mathematics, 95,* 136–146.

Gerard, L. F., Varma, K., Corliss, S. B., & Linn, M. C. (2011). Professional development for technology-enhanced inquiry science. *Review of Educational Research, 81*, 408–448.

Gilbert, J., & Kotelman, M. (2005). 5 good reasons to use notebooks. *Science & Children, 43*(3) 28–32.

Gross, M. U. M. (2006). Exceptionally gifted children: Long-term outcomes of academic acceleration and nonacceleration. *Journal for the Education of the Gifted, 29*, 404–429.

Haley, J., & Monk, L. (2013). *Erosion*. Conway: University of Central Arkansas.

Herron, M. D. (1971). The nature of scientific enquiry. *School Review, 79*(2), 171–212.

International Ocean Discovery Program, The. (2014). *The race is on . . . with seafloor spreading!* Retrieved from http://joidere solution.org/node/3152

Johnsen, S. K., & Sheffield, L. J. (Eds.). (2013). *Using the Common Core State Standards for Mathematics with gifted and advanced learners*. Waco, TX: Prufrock Press.

Kennedy, M. (1998). *Form and substance of inservice teacher education* (Research Monograph No. 13). Madison: University of Wisconsin, National Institute for Science Education.

Krajcik, J., Czerniak, C. M., & Berger, C. F. (2003). *Teaching science in elementary and middle school classrooms: A project–based approach*. Boston, MA: McGraw-Hill.

Little, C. A., & Paul, K. A. (2011). Professional development to support implementation of content-based curriculum. In J. VanTassel-Baska & C. A. Little (Eds.), *Content-based curriculum for high-ability learners* (2nd ed.) (pp. 413–436). Waco, TX: Prufrock Press.

Lumpe, A., Czerniak, C., Haney, J., & Beltyukova, S. (2012). Beliefs about teaching science: The relationship between elementary teachers' participation in professional development and student achievement. *International Journal of Science Education 34*, 153–166.

Manning, M., Szecsi, T., Geiken, R., Van Beeteren, B. D., & Kato, T. (2009). Teaching strategies: Putting the cart before the horse: The role of a socio-moral atmosphere in an inquiry-based curriculum. *Childhood Education, 85*(4), 260–263.

Marcarelli, K. (2010). *Teaching science with interactive notebooks.* Thousand Oaks, CA: Corwin Press.

Marshall, J. C., & Horton, R. M. (2011). The relationship of teacher-facilitated, inquiry-based instruction to student higher-order thinking. *School Science and Mathematics, 111*(3), 93–101.

McIntosh, J. D. (2009). Classroom management, rules, consequences, and rewards! Oh, my! *Science Scope, 32*(9), 49–51.

Melber, L. M. (2003). Partnerships in science learning: Museum outreach and elementary gifted education. *Gifted Child Quarterly, 47,* 251–258.

Michaels, S., Shouse, A. W., & Schweingruber, H. A. (2008). *Ready, set, science! Putting research to work in K–8 science classrooms.* Washington, DC: The National Academies Press.

National Association for Gifted Children. (2004). *Acceleration* [Position Statement]. Retrieved from http://www.nagc. org/about-nagc/who-we-are/nagc-position-statements-white-papers

National Association for Gifted Children. (2008). *The STEM promise: Recognizing and developing talent and expanding opportunities for promising students of science, technology, engineering and mathematics* [White Paper]. Retrieved from http://www. nagc.org/about-nagc/who-we-are/nagc-position-statements-white-papers

National Association for Gifted Children. (2010). *NAGC Pre-K–Grade 12 Gifted Programming Standards: A blueprint for quality gifted education programs.* Washington, DC: Author.

National Research Council. (2007). *Taking science to school: Learning and teaching science in grades K–8.* Washington, DC: The National Academies Press.

National Research Council. (2012). *A Framework for K–12 science education: Practices, crosscutting concepts, and core ideas.* Washington, DC: The National Academies Press.

National Research Council. (2014). *Developing assessments for the Next Generation Science Standards.* Committee on Developing Assessments of Science Proficiency in K–12. Washington, DC: The National Academies Press.

National Science Board. (2010). *Preparing the next generation of STEM innovators: Identifying and developing our nation's human capital.* Retrieved from http://www.nsf.gov/nsb/publica tions/2010/nsb1033.pdf

National Science Teachers Association. (2011). *NSTA position statement: Quality science education and 21st-century skills.* Retrieved from http://www.nsta.org/about/positions/21st century.aspx?print=true

NGSS Lead States. (2013a). *The Next Generation Science Standards: For states, by states, Volume 1, The Standards.* Washington, DC: The National Academies Press.

NGSS Lead States. (2013b). *The Next Generation Science Standards: For states, by states, Volume 2, Appendices.* Washington, DC: The National Academies Press.

NGSS Lead States. (2013c). *The Next Generation Science Standards: Appendix E.* Washington, DC: The National Academies Press. Retrieved from http://www.nextgenscience.org/ sites/ngss/files/Appendix%20E%20-%20Progressions%20 within%20NGSS%20-%20052213.pdf

NGSS Lead States. (2013d). *The Next Generation Science Standards: Appendix F.* Washington, DC: The National Academies Press. Retrieved from http://www.nextgenscience.org/sites/ ngss/files/Appendix%20F%20%20Science%20and%20 Engineering%20Practices%20in%20the%20NGSS%20 -%20FINAL%20060513.pdf

NGSS Lead States. (2013e). *The Next Generation Science Standards: Appendix G.* Washington DC, The National Academies Press. Retrieved from http://www.nextgenscience.org/ sites/ngss/files/Appendix%20G%20-%20Crosscutting%20 Concepts%20FINAL%20edited%204.10.13.pdf

Olszewski–Kubilius, P. (2010). Special schools and other options for gifted STEM students. *Roeper Review, 32,* 61–70.

Partnership for 21st Century Skills. (2009). *P21 framework definitions.* Retrieved from http://www.p21.org/storage/documents/P21_Framework_Definitions.pdf

Penuel, W. R., Gallagher, L. P. & Moorthy, S. (2011). Preparing teachers to design sequences of instruction in earth systems science: A comparison of three professional development programs. *American Educational Research Journal, 48,* 996–1025.

Piaget, J. (1937). *La construction de reel chez l'enfant.* Neuchatel, France: Delachaux et Niestlé.

Poon, C. L., Tan, D., & Tan, A. L. (2009). Classroom management and inquiry-based learning: Finding the balance. *Science Scope, 32*(9), 19–21.

Robinson, A., Dailey, D., Hughes, G., & Cotabish, A. (2014). The effects of a science-focused STEM intervention on gifted elementary students' science knowledge and skills. *Journal of Advanced Academics, 25,* 159 –161.

Roehrig, G. H., & Luft, J. A. (2006). Does one size fit all? The induction experience of beginning science teachers from different teacher-preparation programs. *Journal of Research in Science Teaching, 43*(9), 963–985.

Rogers, K. (2007). Lessons learned about educating the gifted and talented: A synthesis of the research on the educational practice. *Gifted Child Quarterly 51,* 382–396.

Romano, M. (2012). The new teacher's toolbox. *Science Teacher, 079*(7), 14.

Ryser, G. R. (2011). Qualitative and quantitative approaches to assessment. In S. K Johnsen (Ed.), *Identifying gifted students: A practical guide* (pp. 37–62). Waco, TX: Prufrock Press.

Savery, J. R. & Duffy, T. M. (1995). Problem based learning: An instructional model and its constructivist framework. *Educational Technology, 35*(5), 31–37.

Schwartz, M. S., Sadler, P. M., Sonnert, G., & Tai, R. H. (2009). Depth versus breadth: How content coverage in high school science courses relates to later success in college science coursework. *Science Education, 9*(5), 798–826.

Sheffield, L. J. (2006). Developing mathematical promise and creativity. *Journal of the Korea Society of Mathematical Education Series D: Research in Mathematical Education 10*, 1–11.

Siegle, D., & McCoach, D. B. (2002). Promoting a positive achievement attitude with gifted and talented students. In M. Neihart, S. M. Reis, N. M. Robinson, & S. M. Moon (Eds.), *The social and emotional development of gifted children: What do we know?* (pp. 237–249). Waco, TX: Prufock Press.

Sterling, D. R. (2009). Classroom management: Setting up the classroom for learning. *Science Scope, 32*(9), 29–33.

Supovitz, J. A., & Turner, H. M. (2000). The effects of professional development on science teaching practices and classroom culture. *Journal of Research in Science Teaching, 37*, 963–980.

Swafford, J. (1998). Teachers supporting teachers through peer coaching. *Support for Learning, 13*(2), 54–58.

Swiatek, M. A., & Benbow, C. P. (1991). A ten-year longitudinal follow-up of ability matched accelerated and unaccelerated gifted students. *Journal of Educational Psychology, 83*, 528–538.

Swiatek, M. A., & Benbow, C. P. (1992). Nonacademic correlates of satisfaction with accelerative programs. *Journal of Youth and Adolescence, 21*, 699–723.

TeachEngineering. (2013). *Do as the Romans: Construct an aqueduct!* Denver: Regents of the University of Colorado. Retrieved from http://www.teachingengineering.org/view_activity. php?url=collection/wpi_/activities/wpi_construct_an_aque duct/construct_an_aqueduct.xml#mats

Tieso, C. (2005). The effects of grouping practices and curricular adjustments on achievement. *Journal for the Education of the Gifted, 29*(1), 60–89.

Tomlinson, C. A. (2001). *How to differentiate instruction in mixed-ability classrooms*, (2nd ed.). Alexandria, VA: Association for Supervision and Curriculum Development.

Tomlinson, C. A., Kaplan, S. N., Renzulli, J. S., Purcell, J. H., Leppien, J., Burns, D. E., Strickland, C. A., & Imbeau, M. B. (2009). *The parallel curriculum*. Thousand Oaks, CA: Corwin.

United States Geological Survey, Department of the Interior. (2014). *Wegener's puzzling evidence exercise.* Retrieved from http://volcanoes.usgs.gov/about/edu/dynamicplanet/wegener/

VanTassel-Baska, J. L. (Ed.). (2007). *Serving gifted learners beyond the traditional classroom: A guide to alternative programs and services.* Waco, TX: Prufrock Press.

VanTassel-Baska, J., Avery, L. D., Hughes, C. E., & Little, C. A. (2000). An evaluation of the implementation of curriculum innovation: The impact of William and Mary units on schools. *Journal for the Education of the Gifted, 23,* 244–272.

VanTassel-Baska, J., & Bass, G. (1998). A national study of science curriculum effectiveness with high ability learners. *Gifted Child Quarterly, 42,* 200–211.

Vygotsky, L. S. (1978). *Mind in society: The development of higher psychological processes.* Boston, MA: Harvard University Press.

Williams, L. M., Brittingham, M. C., & Smith, S. S. (2001). *The wildlife ecologist.* University Park: Pennsylvania State University.

Wojnowski, B. S., & Pea, C. H. (2014). *Models and approaches to STEM professional development.* Arlington, VA: National Science Teachers Association.

Wolfgang, C. N. (2009). Managing inquiry-based classrooms. *Science Scope, 32*(9), 15–17.

Wood, S. (2002). Perspectives of best practices for learning gender-inclusive science: Influences of extracurricular science for gifted girls and electrical engineering for women. *Journal of Women and Minorities in Science and Engineering, 8,* 25–40.

Yoon, K. S., Duncan, T., Lee, S., & Shapley, K. (2008, March). *The effects of teachers' professional development on student achievement: Findings from a systemic review of evidence.* Paper presented at the annual meeting of the American Educational Research Association: New York, NY.

Appendix A

Definitions of Key Terms

Above-level testing occurs when a student is assessed with a version of a test intended for older students or students who are in grade levels above the assessed student's current grade level. This testing compensates for the fact that many tests (particularly achievement tests) have ceiling effects. In other words, there are not enough difficult items on the test. Tests that are grade calibrated are usually too easy for gifted students, and above-level testing allows educators to test a student's limits or to measure adequately the extent of the gifted student's knowledge.

Acceleration is a broad term used to describe ways in which gifted student learning may occur at a fast and appropriate rate throughout the years of schooling. It refers to content acceleration through preassessment, compacting, reorganizing curriculum by unit or year, grade skipping, telescoping 2 years into one, dual enrollment in high school and college or university, as well as more personalized approaches such as tutorials, mentorships, and independent research that also would be sensitive to the advanced starting level of these learners for instruction. Both Advanced Placement (AP) courses and International Baccalaureate (IB) programs at the high school level are already accelerated in content.

AP courses also may be taken on a fast-track schedule earlier as appropriate.

Appropriate pacing refers to the rate at which material is taught to advanced learners. Because they are often capable of mastering new material more rapidly than typical learners, appropriate pacing would involve careful preassessment to determine readiness for more advanced material to ensure that advanced learners are not bored with the material and are being adequately challenged. Note that although students might advance quickly through some material, they should also be given time to delve more deeply into topics of interest at appropriately advanced levels of complexity and innovation.

Assessment is the way to determine the scope and degree of learning that has been mastered by the student. For purposes of gifted education, the assessments must be matched to differentiated outcomes, requiring the use of authentic approaches like performance- and portfolio-based assessment demands. Some assessments are already constructed and available for use, exhibiting strong technical adequacy and employed in research studies, although others may be teacher developed, with opportunities to establish interrater reliability among teachers who may be using them in schools. Care should be taken to use assessments that do not restrict the level of proficiency that students can demonstrate, such as above-grade assessments that allow for innovative and more complex responses.

Characteristics and needs of gifted learners are the basis for differentiating any curriculum area. Scientifically talented learners often have strong spatial skills, see relationships, recognize patterns, make generalizations, and may be highly fluent, flexible, and original at problem finding and scientific inquiry at an earlier stage of development than typical learners. Because of this advanced readiness, these students may need to be accelerated through the basic material in science in order to focus on higher level science concepts and problems.

Complexity refers to a feature of differentiation that provides advanced learners more variables to study, asks them to use multiple resources to solve a problem, or requires them to use multiple higher order skills simultaneously. The degree of complexity may depend on the developmental level of the learner, the nature of the learning task, and the readiness to take on the level of challenge required.

Creativity and innovation are used to suggest that activities used with the gifted employ opportunities for more open-ended project work that mirrors real-world professional work in solving problems in the disciplines. The terms also suggest that advanced learners are proficient in the skills and habits of mind associated with being a creator or innovator in a chosen field of endeavor. Thus, creative thinking and problem-solving skills would be emphasized.

Curriculum is a set of planned learning experiences, delineated from a framework of expectations at the goal or outcome level that represents important knowledge, skills, and concepts to be learned. Differentiated curriculum units of study already have been designed and tested for effectiveness in science, or units may be developed by teachers to use in gifted instruction.

Differentiation of curriculum for gifted learners is the process of adapting and modifying curriculum structures to address these characteristics and needs more optimally. Thus, curriculum goals, outcomes, and activities may be tailored for gifted learners to accommodate their needs. Typically, this process involves the use of the strategies of acceleration, complexity, depth, and creativity in combination.

Instruction is the delivery system for teaching that comprises the deliberate use of models, strategies, and supportive management techniques. For gifted learners, inquiry strategies such as problem-based learning and creative problem solving and problem posing, and critical-thinking models such as Paul's Reasoning Model, used in independent research or within a flexible grouping approach in the regular classroom constitute instructional differentiation.

Learning progressions are designed to help teachers identify what is expected from their students and the key points along paths that indicate growth in a student's knowledge and skills. As such, they articulate the essential core concepts and processes in each science domain within-and across grade levels.

Rigor and relevance suggest that the curriculum experiences planned for advanced learners be sufficiently challenging yet provided in real-world or curricular contexts that matter to learners at their particular stage of development.

Talent trajectory is used to describe the school span development of advanced learners in their area of greatest aptitude from K–12. It is linked to developmental stages from early childhood through adolescence and defines key interventions that aid in the talent development process, specific to the subject area and desired career path.

Teacher quality refers to the movement at all levels of education to improve the knowledge base and skills of classroom teachers at P–12 levels, which is necessary for effective instruction for advanced students. It is the basis for a redesign of teacher education standards and a rationale for examining P–12 student outcomes in judging the efficacy of higher education programs for teachers. Policy makers are committed to this issue in improving our P–16 education program.

Appendix B

Critical Readings

Adams, C., & Callahan, C. M. (1995). The reliability and validity of a performance task for evaluating science process skills. *Gifted Child Quarterly, 39*, 14–20.

Summary: This paper reports on a study to determine the reliability of an instrument to identify students with aptitude in science. The areas addressed in this study were: interrater reliability, intrarater reliability, equivalent forms reliability, and stability. A sample was taken of 176 students from five school districts. The reliability of the instrument was sufficiently high to warrant validity studies.

Adams. C., Cotabish, A., & Ricci, M. K. (2014). *Using the next generation science standards with advanced and gifted Learners.* Waco, TX: Prufrock Press.

Summary: This book provides teachers and administrators examples and strategies to implement the Next Generation Science Standards (NGSS) with gifted and advanced learners at all stages of development in K–12 schools. The book describes—and demonstrates with specific examples from the NGSS—what effective differentiated activities in science look like for high-ability learners. It shares how educators can provide rigor within the new standards to allow students to demonstrate higher level thinking, reasoning, problem solving, passion, and inventiveness in science. By doing so, students will develop the skills, habits of mind, and attitudes toward learning needed to reach high levels of competency and creative production in science fields.

Akinoglu, O., & Tandogan, R. O. (2007). The effects of problem-based active learning in science education on students' academic achievement, attitude, and concept learning. *Eurasia Journal of Mathematics, Science and Technology Education, 3*(1), 71–81.

Summary: The aim of this study was to determine the effects of problem-based active learning in science education on students' academic achievement and concept learning. The research study was conducted on 50 public school students in seventh grade in the 2004–2005 school year in Istanbul. The treatment process took 30 class hours in total. Three measurement instruments were used: an achievement test, open-ended questions, and an attitude scale for science education. Although the subject matters were taught on the basis of problem-based active learning in the treatment group, traditional teaching methods were employed in the control group. Based on the data collected and the evaluations made in the research, it was determined that the implementation of a problem-based active learning model had positively affected students' academic achievement and their attitudes toward the science course. It was also found that the application of a problem-based active learning model affects students' conceptual development positively and keeps their misconceptions at the lowest level.

Appleton, K. (2008). Developing science pedagogical content knowledge through mentoring elementary teachers. *Journal of Science Teacher Education, 19*, 523–545.

Summary: This study constitutes two case studies of a professional development program for elementary teachers involving mentoring by a university professor. The mentor took the role of a critical friend in joint planning and teaching of science. The study examines the nature of the mentoring relationship and reports the type of teaching learning that occurred, with a particular focus on the teachers' development of science knowledge and skills.

Brandwein, P. F. (1995). *Science talent in the young expressed within ecologies of achievement* (RBDM 9510). Storrs: The University of Connecticut, National Research Center on the Gifted and Talented.
Summary: Six interrelated constructs form the body of this study. The first describes a skein of achievement-centered, goal-targeted environments that do—or should—comprise the inspiring teaching and learning that can enhance the endowments of the young. The second presents studies of unfavorable environments that block the goals of equal opportunity, optimum achievement in science, and the discovery of science proneness or talent. The third comprises elements of formal learning in augmenting environments focusing on instruction as an event evoking early discovery through self-identification of gifted children with a particular bent (or proneness) to science. The fourth is based in the conviction that curriculum and instruction are distinct but related fields within present models of instructed learning. The fifth exemplifies curriculum and instruction, focused in special aptitudes and abilities, relevant to science proneness as a precursor to self-identification of a science talent. The sixth concerns science talent in practice. It describes a skein of discoveries, one leading to another, and concludes with a definition of science talent.

Brody, L. (Ed.). (2004). *Grouping and acceleration practices in gifted and education* (Vol. 3). Thousand Oaks, CA: Corwin.
Summary: This volume of seminal articles from *Gifted Child Quarterly* on grouping and acceleration emphasizes the importance of flexibility when assigning students to instructional

groups, modifying the groups when necessary. Grouping and acceleration have proved to be viable tools to differentiate content for students with different learning needs based on cognitive abilities and achievement levels.

Colangelo, N., Assouline, S. G., & Gross, M. U. M. (Eds.). (2004). *A nation deceived: How schools hold back America's brightest students* (Vol. 2). Iowa City: The University of Iowa, The Connie Belin & Jacqueline N. Blank International Center for Gifted Education and Talent Development.
Summary: Volume 2 of the international report on academic acceleration presents supporting evidence for 18 forms of acceleration and its long-term effects. Interviewed years later, an overwhelming majority of accelerated students say that acceleration was an excellent experience for them and they wish they had accelerated sooner and had more accelerated experiences. They feel academically challenged and socially accepted, and they do not fall prey to the boredom that plagues many highly capable students who are forced to follow the curriculum for their age-peers.

Cotabish, A., Dailey, D., Hughes, A., & Robinson, A. (2011). The effects of a STEM professional development intervention on elementary teachers' science process skills. *Research in the Schools, 18*(2), 16–25.
Summary: This study was part of a larger randomized field study of teacher and student learning in science. This manuscript reports the effects of 2 years of sustained, embedded professional development on grade 2 through 5 elementary teachers' science process skills, as defined by their ability to design experiments, and teacher perceptions of their capacity to lead students in scientific explorations. The results revealed a statistically significant gain in teachers' science process skills, and increased teacher perceptions about their own science process skills as well as their students' science process skills.

Cotabish, A., Dailey, D., Robinson, A., & Hughes, G. (2013). The effects of a STEM intervention on elementary students' science knowledge and skills. *School Science and Mathematics, 113*(5), 215–226.

Summary: The purpose of the study was to assess elementary students' science process skills, content knowledge, and concept knowledge after one year of participation in an elementary STEM program. This study documented the effects of the combination of intensive professional development and the use of inquiry-based science instruction in the elementary classroom, including the benefits of using rigorous science curriculum with general education students. The results of the study revealed a statistically significant gain in science process skills, science concepts, and science content knowledge by students in the experimental group when compared with students in the comparison group. Moreover, teacher participation in the STEM professional development program had a statistically significant impact on students' variability in posttest scores.

Dailey, D., Cotabish, A., & Robinson, A. (2013). A model for STEM talent development: Peer coaching in the elementary science classroom. *TEMPO, 4*(33), 15–19.

Summary: This manuscript provides details about the Javits grant-funded *STEM Starters* peer coaching model as well as teacher professional development focused on science.

Johnsen, S. (2005). Within-class acceleration. *Gifted Child Today, 28*(1), 5.

Summary: This article describes ways teachers can accelerate the curriculum in their classroom by preassessing students and modifying their instruction, allowing them either to move through the curriculum at a faster pace or to provide in-depth learning experiences.

National Academy of Sciences. (2007). *Rising above the gathering storm: Energizing and employing America for a brighter economic future.* Retrieved from www.utsystem.edu/competitive/files/RAGS-fullreport.pdf
Summary: This seminal report describes the erosion of U.S. advantages in the marketplace and in science and technology. The report states that a comprehensive and coordinated federal effort is urgently needed to bolster U.S. competitiveness and pre-eminence in these areas so that the nation will consistently gain from the opportunities offered by rapid globalization. Four recommendations were made that focus on actions in K–12 education, higher education, and economic policy.

National Academy of Sciences. (2010). *Rising above the gathering storm, revisited: Rapidly approaching category 5.* Washington, DC: The National Academies Press.
Summary: As a follow-up publication to *Rising Above the Gathering Storm* (2007), *Rising Above the Gathering Storm, Revisited* provides a snapshot of the work of the government and the private sector in the intervening 5 years, analyzing how the original recommendations have or have not been acted upon, what consequences this may have on future competitiveness, and priorities going forward. In addition, readers will find a series of thought- and discussion-provoking factoids—many of them alarming—about the state of science and innovation in America.

National Association for Gifted Children, Task Force on Math and Science. (2008). *The STEM promise: Recognizing and developing talent and expanding opportunities for promising students of science, technology, engineering and mathematics.* Retrieved from http://www.nagc.org/about-nagc/who-we-are/nagc-position-statements-white-papers
Summary: This White Paper draws attention to the pervasive need to develop and engage mathematically and scientifically advanced learners. Research and implications are included.

National Science Board. (2010). *Preparing the next generation of STEM innovators: Identifying and developing our nation's human capital* (NSB–10–33). Retrieved from http://www.nsf.gov/nsb/publications/2010/nsb1033.pdf

Summary: In this report, the National Science Board addresses talent identification and development of children and young adults, and provides recommendations that enhance the STEM innovation capacity of the United States.

National Science Teachers Association. (2011). *NSTA position statement: Quality science education and 21st-century skills.* Retrieved from http://www.nsta.org/about/positions/21st century.aspx?print=true

Summary: This NSTA position paper presents declarations that are in concert with other position statements published by NSTA that outline goals for quality science education. Specifically, the paper makes recommendations to support 21st-century skills consistent with best practices across a science education system, including curriculum, pedagogy, science teacher preparation, and teacher professional development.

NGSS Lead States. (2013). *Next Generation Science Standards: For states, by states, Volume 1, The Standards.* Washington, DC: The National Academies Press.

Summary: Through a collaborative, state-led process managed by Achieve, Inc., K–12 standards have been developed that are rich in content and practice, arranged in a coherent manner across disciplines and grades to provide all students an internationally benchmarked science education. The NGSS are based on the Framework for K–12 Science Education developed by the National Research Council.

Olszewski-Kubilius, P. (2010). Special schools and other options for gifted STEM students. *Roeper Review, 32,* 61–70.

Summary: In this article, the author compares the advantages and disadvantages of special STEM schools to other options such as summer programs, distance learning, mentorships, and others, as well as discusses the characteristics of students for whom each option is most appropriate. Alternatives to STEM schools, including piecing together summer programs, distance-learning programs, and other options, are also discussed and evaluated.

Park, G., Lubinski, D., & Benbow, C. P. (2012). When less is more: Effects of grade skipping on adult STEM productivity among mathematically precocious adolescents. *American Psychological Association, 105*(1), 176–198.

Summary: Using data from a 40-year longitudinal study, the authors examined three related hypotheses about the effects of grade skipping on future educational and occupational outcomes in science, technology, engineering, and mathematics (STEM). From a combined sample of 3,467 mathematically precocious students (top 1%), a combination of exact and propensity score matching was used to create balanced comparison groups of 363 grade skippers and 657 matched controls. Results suggest that grade skippers (a) were more likely to pursue advanced degrees in STEM and peer-reviewed publications in STEM, (b) earned their degrees and authored their first publication earlier, and (c) accrued more total citations and highly cited publications by the age of 50. The patterns were consistent among male participants but less so among female participants (who had a greater tendency to pursue advanced degrees in medicine or law). Findings suggest that grade skipping may enhance STEM accomplishments among the mathematically talented.

Reis, S. M., & Park, S. (2001). Gender differences in high-achieving students in math and science. *Journal for the Education of the Gifted, 25*(1), 52–73.

Summary: Using data from the National Education Longitudinal Study of 1988, the researchers examined gender differences between high-achieving students in math and science. They found that there were more high-achieving males than females in this group, with far fewer female students in the science group. They also found that high-achieving males felt better about themselves than high-achieving females. Females who are high-achieving in math and science are more influenced than males are by teachers and families.

Robbins, J. I. (2011). Adapting science curricula for high-ability learners. In J. VanTassel-Baska & C. A. Little (Eds.), *Content-based curriculum for high-ability learners* (2nd ed., pp. 437–465). Waco, TX: Prufrock Press.
Summary: This chapter focuses on adapting science curricula for high-ability learners. It is intended to facilitate differentiated curriculum development for gifted students that is substantive, rigorous, and in keeping with disciplinary structures.

Roberts, J. L. (2010). Talent development in STEM disciplines: Diversity—cast a wide net. *National Consortium for Specialized Secondary Schools of Mathematics, Science, and Technology, 16*(1), 10–12. Retrieved from http://www.eric.ed.gov/PDFS/EJ930651.pdf
Summary: In this column, the author addresses STEM talent development among diverse learners, recommends steps necessary to recognize potential talent, and provides recommendations for appropriately challenging learning opportunities to develop them to optimum levels.

Robinson, A., Shore, B. M., & Enersen, D. L. (2007). *Best practices in gifted education: An evidence-based guide.* Waco, TX: Prufrock Press.
Summary: This book provides concise, research-based advice to educators, administrators, and parents of gifted and talented youth. The 29 practices included in this volume are the result of an extensive examination of educational research on what works with talented youth. The interest in culturally diverse and low-income learners, the means to identify talents, and the need for curriculum that appropriately challenges high-ability youth constitute just a few of the 29 practices.

Van Tassel-Baska, J. (Ed.). (2004). *Curriculum for gifted and talented students* (Vol. 4). Thousand Oaks, CA: Corwin Press.

Summary: A collection of seminal articles and research from *Gifted Child Quarterly* are compiled in one volume, including articles on how to develop a scope and sequence for the gifted, the multiple menu model of serving gifted students, what effective curriculum for the gifted looks like, curriculum at the secondary level, and specific content-area curricula options in math and science.

VanTassel-Baska, J. (1998). Planning science programs for high ability learners. (ERIC Digest No. E546). Retrieved from http://www.ericdigests.org/1999-3/science.htm

Summary: This manuscript addresses the research of gifted learners in science, provides recommendation for science curriculum reform including recommendations for teachers, and includes a curriculum reform classroom indicators checklist.

VanTassel-Baska, J., Bass, G., Ries, R., Poland, D., & Avery, L. (1998). A national study of science curriculum effectiveness with high ability students. *Gifted Child Quarterly, 42*(4), 200–211.

Summary: This study assessed student growth on integrated science process skills after being taught a 20–36-hour science unit. The prototypical unit, "Acid, Acid Everywhere," was implemented in 15 school districts across seven states. The unit was based on the Integrated Curriculum Model (ICM), developed specifically for gifted learners, which stresses advanced content, high-level process and product, and a concept dimension. Results indicate small, but significant, gains for students in integrated science process skills when compared to equally able students not using the units. Implementation data reflected satisfaction of teachers with the units, especially in terms of student interest and motivation. The effectiveness of this curriculum, designed to align with the new science standards and to be appropriate for gifted students, lends credibility to the argument for using the new content standards as a basis for curriculum development efforts with gifted learners.

VanTassel-Baska, J., & Stambaugh, T. (Eds.). (2008). What works: Twenty years of curriculum development and research for advanced learners. Retrieved from http://education.wm.edu/centers/cfge/curriculum/documents/WhatWorks.pdf

Summary: This document highlights "what works" based on the curriculum development and research work of the William and Mary Center for Gifted Education during a 20-year span. Areas of study include curriculum development, instruction, assessment, and professional development. Using the Integrated Curriculum Model (ICM) as a template for design, coupled with curriculum reform emphases in content areas including science, the Center curriculum produced positive outcomes in student achievement and teacher use of differentiated instruction.

Teacher Resources

Books

Adams, C. M., Cotabish, A., & Ricci, M. A. (2014). *Using the Next Generation Science Standards with gifted and advanced learners.* Waco, TX: Prufrock Press.

Description: The book assists educators in differentiating the NGSS for advanced learners. It presents common lessons targeting each disciplinary core idea (elementary, middle, and high school) and provides alternative lessons for advanced learners. The book also envisions how the NGSS can be integrated with gifted education curricula, instructional practices, and program models.

Bybee, R. W. (2013). *Translating the NGSS for classroom instruction.* Arlington, VA: National Science Teachers Association.

Description: This book guides the implementation of the NGSS. Using the BSCS 5E Instructional Model, Bybee provides examples on how to translate performance expectations from the standards into classroom practices in elementary, middle, and high school. Additionally, he assists educators with aligning current units of instruction to the standards and provides examples of curriculum, instruction, and assessment. Bybee also shares insights with teachers, school administrators, and state science coordinators on reforming science curriculum to address the standards.

National Research Council. (2012). *A framework for K–12 science education: Practices, crosscutting concepts, and core ideas.* Washington, DC: The National Academies Press.
Description: The *Framework* is the foundation for the NGSS. The book presents readers with the new vision for K–12 science education that provided the direction for the NGSS. Drawing on current research, the *Framework* identifies the key scientific knowledge and skills all students should learn by the end of high school. To increase student understanding of science over time, the *Framework* recommends classroom instruction involve the integration of three dimensions: Scientific and Engineering Practices, Crosscutting Concepts, and Disciplinary Core Ideas.

Pratt, H. (2013). *The NSTA reader's guide to the Next Generation Science Standards.* Arlington, VA: National Science Teachers Association.
Description: This book offers readers assistance in understanding the NGSS and provides planning tools and resources to assist in implementing the standards.

Curriculum

Blueprints for Biography®
Description: STEM Starters, a Javits grant project, developed literacy-focused science curriculum guides to be used with gifted learners. *Blueprints for Biography*® is a series of teacher curriculum guides with high-level discussion questions, creative and critical thinking activities, a persuasive writing component, and rich primary resources. STEM *Blueprints* focus on eminent scientists and inventors for whom exemplary children's biographies exist in trade book form. Paper format *Blueprints* currently exist for the following: George Washington Carver, Galileo Galilei, Thomas

Edison, Marie Curie, Alexander Graham Bell, Michael Faraday, Louis Pasteur, and Albert Einstein. Each guide concludes with a classic experiment for students to carry out. An innovative peer coaching model specific to science instruction was also developed. For more information about STEM Starters, visit http://ualr.edu/gifted/index.php/home/stem/.

The Center for Gifted Education at the College of William and Mary Science Curriculum Units

Description: The Center for Gifted Education at the College of William and Mary has designed curricular units in the areas of mathematics, language arts, science, and social studies that are based on the three dimensions of the Integrated Curriculum Model: advanced content, higher level processes and products, and interdisciplinary concepts, issues, and themes. The materials emphasize a sophistication of ideas, opportunities for extensions, the use of higher order thinking skills, and opportunities for student exploration based on interest. The science curriculum units are geared toward different grade level clusters, yet can be adapted for use at all levels K–8. Through these units, students experience the work of real science in applying data-handling skills, analyzing information, evaluating results, and learning to communicate their understanding to others. For more information about the William and Mary units, visit the Center for Gifted Education at http://education.wm.edu/centers/cfge/curriculum/index.php.

Engineering is Elementary

Description: With the inclusion of engineering in the NGSS, it becomes increasingly important to provide teachers with quality resources to assist them in meeting the engineering standards. *Engineering is Elementary* offers multiple curricular units in all domains of science for grades 1–8. The units include a teacher's guide, storybook, and materials list. In addition, multiple free lessons are found on their website. For more information, visit http://www.eie.org/about-us.

Seeds of Science/Roots of Reading

Description: *Seeds of Science/Roots of Reading* was developed for students in grades 2–5 and designed to integrate science and literacy. The curriculum provides a focus on essential science understandings while building a full range of literacy skills. Using rigorous field testing, researchers found students using Seeds of Science/Roots of Reading demonstrated increased student achievement in both literacy and science. For more information, visit http://www.scienceandliteracy.org/.

U–STARS˜Plus

Description: *U–STARS˜Plus*, a Javits grant project, provides Science and Literature Connections to explore scientific ideas within literacy instruction time using 32 popular children's books. Science and Literature Connections is organized around Bloom's taxonomy to support a range of thinking levels and to scaffold learning. By using these materials, a teacher can create a higher level thinking environment around literature connected with science, which motivates reluctant readers. For more information about U–STARS˜Plus Science and Literature Connection, visit http://www.fpg.unc.edu/node/4010.

Websites

https://cogito.org/

Description: Cogito.org was developed by the Johns Hopkins Center for Talented Youth (CTY) in collaboration with other talent search centers. The goals are to address the needs of gifted students around the world with abilities and interests in math and science. The public site is full of news, interviews with scientists, profiles of young scientists, and a searchable database of programs. Students invited to join as members become part of an online community and participate in discussion forums with each other and with experts in their fields.

http://www.davidsongifted.org/db/browse_by_topic_resources.aspx

Description: The Davidson Institute for Talent Development offers links to resources in mathematics, language arts, science, social studies, arts and culture, and related domains. It also provides links to information about educational options such as ability grouping, acceleration, enrichment programs, competitions, and other services.

http://www.gifted.uconn.edu/nrcgt/nrconlin.html

Description: The Neag Center for Gifted Education and Talent Development offers online resources that describe research studies and defensible practices in the field of gifted and talented education. Some of the studies address curriculum at the high school level, the explicit teaching of thinking skills, cluster grouping, algebraic understanding, reading with young children, differentiated performance assessments, and content-based curriculum.

http://www.lawrencehallofscience.org/comsci/

Description: Lawrence Hall's Communicating Science course is a teacher education course that engages prospective educators in learner-centered practices with a strong focus on inquiry-based science. On this website, there are free resources from the Communicating Science course, including PowerPoints and handouts, for use by multiple audiences and in various educational settings.

http://ngss.nsta.org/

Description: The National Science Teachers Association (NSTA) has curricula, assessment, and instructional resources for implementing the NGSS for all students, along with recommendations for a research agenda related to the Standards.

http://www.pearsonschool.com/index.cfm?locator=PS1p1u

Description: Pearson provides free resources to help educators implement the NGSS. The website offers example lessons for elementary, middle, and high school teachers and they present free online training videos for administrators, teachers and parents.

http://serendip.brynmawr.edu/sci_edu/waldron/
Description: This website at Bryn Mawr University provides hands-on/minds-on activities for teaching biology to middle and high school students. Some of the activities have been linked to the NGSS.

http://serendip.brynmawr.edu/exchange/bioactivities/NGSS
Description: This website hosted by Bryn Mawr University links all the activities at the Hands-On Activities website (see above) to their appropriate NGSS performance expectations.

http://http://joidesresolution.org/
Description: The International Ocean Discovery Program offers teachers and students resources to investigate oceanic science.

http://www.nasa.gov/audience/foreducators/index.html#.U20vXfldUiA
Description: NASA provides a variety of tools, activities, and resources for teachers and students interested in STEM.

http://studentclimatedata.unh.edu/index.shtml
Description: The University of New Hampshire provides a website designed to engage students in the study of climate change. The website provides teachers lesson plans, resources, and student materials. Students and teachers have access to climate change data and tools that can be used in observing past, current, and future climatic trends.

http://www.nws.noaa.gov/om/edures.shtml
Description: The National Weather Service website offers educators materials that are useful in the investigation of weather and

weather related invents. The website also contains fun activities and games for students.

Talent Searches

Talent searches offer gifted students opportunities to assess their talent, identify their strengths, and explore their educational options. Many programs provide summer-type camps or online access where students can study advanced content and concepts, interact with students of similar abilities and interests, and receive recognition for academic achievement. The list below describes several of the more well-known talent searches. For a comprehensive list, see the Davidson Institute for Talent Development at http://www.davidsongifted.org/db/browse_resources_129.aspx.

Name: Academic Talent Search (ATS); University of California, Irvine
Website: http://www.giftedstudents.uci.edu/ats/
Description: ATS is a testing program primarily for student in grades 6–10. It identifies students with extraordinary mathematical and/or verbal reasoning abilities and offers students an opportunity to improve test-taking skills for the PSAT and SAT Reasoning Test.

Name: Belin-Blank Exceptional Student Talent Search (BESTS); University of Iowa
Website: http://www.education.uiowa.edu/belinblank/Students/BESTS/
Description: BESTS uses above-level testing to identify students in grades 4–9 who need further educational challenge to fully realize their academic talent. BESTS students are eligible to participate in Belin-Blank Center precollege programs consisting of a range of summer opportunities as well as an academic-year program.

Name: The Carnegie Mellon Institute for Talented Elementary and Secondary Students (C-MITES)
Website: http://www.cmu.edu/cmites/index.html
Description: C-MITES provides academic summer and weekend programs to gifted students in kindergarten through 10th grade. The hands-on workshops emphasize exploration and investigation of advanced topics. They bring talented students together in a group, thus supporting their social and emotional needs and building a community of young scholars. Classes include Build a Robot, Maglev, Amusement Park Physics, Programming Using Alice, Creative Writing, and Informal Geometry. C-MITES offers above-level testing for third through sixth graders, professional development for teachers, and informational workshops and publications for parents.

Name: Center for Bright Kids (CBK)
Website: http://www.centerforbrightkids.org
Description: CBK is a regional talent center located in Westminster, CO. In addition to the traditional talent search testing, CBK offers school- and district-specific trainings on partnerships to use above-level testing scores as an assessment tool for programs, as well as evaluation as part of an accelerated learning program. The CBK testing interpretation guide includes an addendum on "what to do next," with various models of talent development offered as a way to understand how to utilize scores.

Name: Center for Talented Youth (CTY); John Hopkins University
Website: http://cty.jhu.edu/
Description: CTY's talent search identifies advanced learners in grades 2–8 and serves as the bridge to CTY, including its summer programs and online courses. CTY offers students greater academic challenges, interaction with intellectual peers, and teaching strategies designed especially for the gifted.

Name: Northwestern University's Midwest Academic Talent Search (NUMATS).
Website: http://www.ctd.northwestern.edu/numats/
Description: NUMATS identifies talent through above-grade-level testing and serves students in grades 3–9 in Indiana, Michigan, Minnesota, Ohio, Wisconsin, Illinois, North Dakota, and South Dakota. NUMATS provides identified students with appropriately challenging programs and resources and serves as a gateway to programs and resources for gifted students both at Northwestern University and nationwide. Families and educators participating in NUMATS gain access to an online, password-protected toolbox. Through the family toolbox, students are offered invaluable test preparation materials, test result analysis, in-depth articles, webinars, and continuously updated resources specific to understanding and developing the academic talent of their gifted child. Through the educator toolbox, schools with participating students receive face-to-face and online professional development and consultation.

Name: Talent Identification Program (TIP); Duke University
Website: http://www.tip.duke.edu/
Description: The Duke University Talent Identification Program (Duke TIP) identifies gifted children in fourth through seventh grades using above-grade testing and provides resources to nurture the development of these exceptionally bright youngsters. Duke TIP offers several resources to help participants use their above-level test results as a guide to maximize their education. After testing, participants receive a results summary that compares their performance to other talent search participants and offers suggestions for educational development. TIP's wide variety of educational programs, both residential and commuter, engages students through superb academic experiences. The Educational Opportunity Guide, an online directory, contains listings for schools, summer programs, and academic competitions across the United States and abroad, and the *Digest of Gifted Research* offers research-based information about raising and educating academically talented children.

About the Authors

Cheryll M. Adams, Ph.D., is the Director Emerita of the Center for Gifted Studies and Talent Development at Ball State University and teaches graduate courses in gifted education. She has authored or coauthored numerous publications in professional journals, as well as several books and book chapters. She has also coauthored and directed three federal Javits grants. She serves on the editorial review board for several journals. She has served on the Board of Directors of the NAGC and has been president of the Indiana Association for the Gifted and of The Association for the Gifted, Council for Exceptional Children. She was a science teacher and gifted education coordinator for 15 years.

Alicia Cotabish, Ed.D., is an assistant professor of Teaching and Learning at the University of Central Arkansas. Previously, she was the Associate Director of the Jodie Mahony Center for Gifted Education and Advanced Placement Professional Development Center at the University of Arkansas at Little Rock, and served as one of two Principal Investigators and the Director of *STEM Starters*, a federally funded Javits project. From 2004 to 2007, Dr. Cotabish coordinated the *Arkansas Evaluation Initiative in Gifted Education*, a Jacob K. Javits-funded statewide school district program evaluation initiative housed at the University of Arkansas at Little Rock. Before beginning her career in higher education in 2004, Dr. Cotabish taught for 8 years in the public school system as an elementary and middle school science teacher, a K–12 gifted and talented teacher, and as a gifted education program administrator.

Debbie Dailey, Ed.D., is an assistant professor of teaching and learning at the University of Central Arkansas. Formerly, Debbie was the Associate Director for the Jodie Mahony Center for Gifted Education at the University of Arkansas at Little Rock. Debbie also served as the Curriculum Coordinator and Peer Coach for a Javits-funded program, *STEM Starters*, which focused on improving science instruction in the elementary grades. Prior to moving to higher education, Debbie was a high school science teacher and gifted education teacher for 20 years.

Printed in the United States
by Baker & Taylor Publisher Services